S0-CAH-058

The Structure of Rare-Earth Metal Surfaces

The Structure of Rare-Earth Metal Surfaces

S. D. Barrett

University of Liverpool, UK

S. S. Dhesi

European Synchrotron Radiation Facility, France

ICP

Imperial College Press

Published by

Imperial College Press
57 Shelton Street
Covent Garden
London WC2H 9HE

Distributed by

World Scientific Publishing Co. Pte. Ltd.

P O Box 128, Farrer Road, Singapore 912805

USA office: Suite 1B, 1060 Main Street, River Edge, NJ 07661

UK office: 57 Shelton Street, Covent Garden, London WC2H 9HE

British Library Cataloguing-in-Publication Data
A catalogue record for this book is available from the British Library.

THE STRUCTURE OF RARE-EARTH METAL SURFACES

ISBN 1-86094-165-6

Printed in Singapore by Uto-Print

This book is dedicated to the graduates of the
Rare Earth Group 1988–1999

Rob, Adam, Martin, Richard,
Myoung, Nigel and Chris

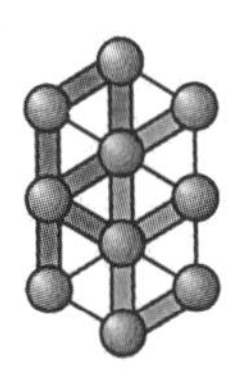

and to Robin and Tony

CONTENTS

PREFACE

The Structure of Rare–Earth Metal Surfaces outlines the experimental techniques and theoretical calculations employed in the study of the atomic and nanoscale structure of surfaces and reviews their application to the surfaces of the rare–earth metals. In particular, the results of quantitative low–energy electron diffraction experiments and multiple–scattering calculations are covered in some detail due to their importance in the field of surface structure determination.

The aim of the book is two–fold. Firstly, it can be used by those new to the field of surface science as an introduction to surface crystallography, covering the basic principles and showing many examples of surfaces from a selection of elements from the periodic table. Secondly, it can be used as an introduction to the world of the rare–earth metals, and their surfaces in particular, giving a taste of the active research fields that are currently being explored. A full review of all of the many hundreds of studies that have been made of rare–earth metal surfaces is not attempted, as it would be too large a task and beyond the scope of this book. However, we thought that the information tabulated in the appendix, giving references for all studies that have some relevance to rare–earth metal surfaces, would be a valuable source of information for any student embarking on research in this or related fields.

Surface science is, by its very nature, an interdisciplinary science — often no clear boundaries exist between the physics, chemistry and materials science of surfaces. Any that do appear to exist are either imposed artificially or are very ill–defined, or perhaps both. This book may appear to fall into the category of surface physics, as it describes the principles and practice of determining the crystallographic structure of surfaces at the atomic and nanometre scales. However, by describing how the structures of surfaces are intimately linked to their electronic, magnetic and chemical properties, we show that surface structure cannot be studied in total isolation from these other aspects. We have aimed the book at science graduates with an interest in surface crystallography. Although a background in solid state physics will be helpful to the reader, by not relying too heavily on undergraduate physics we have tried to keep the content at a level such that graduates from disciplines other than physics will not be disadvantaged. Describing the theoretical treatment of surface atomic structure determination will inevitably require some mathematics, but where this is unavoidable the reader is led through the mathematics to see how the physics comes through, sacrificing rigour for clarity where possible.

We cannot give a comprehensive account of all aspects of either surface crystallography or the rare–earth metals, and so further reading paragraphs are provided at the end of some of the sections to guide the reader to review articles or books that give more detailed information on techniques, applications, or related studies.

Steve Barrett
Sarnjeet Dhesi

LIST OF ACRONYMS

AES	Auger electron spectroscopy
ARUPS	angle–resolved ultraviolet photoelectron spectroscopy
bcc	body–centred cubic
CLS	core–level shift
CTR	crystal truncation rod
dhcp	double hexagonal close–packed
fcc	face–centred cubic
FEM	field emission microscopy (or field electron microscopy)
FM	Frank–van der Merwe
hcp	hexagonal close–packed
IEC	interface energy calculation
I–V	intensity–voltage (LEED) or current–voltage (STM)
KS	Kurdjumov–Sachs
LD	layer doubling
LECBD	low–energy cluster beam deposition
LEED	low–energy electron diffraction
LSDA	local–spin–density approximation
MBE	molecular beam epitaxy
MEIS	medium–energy ion scattering
ML	monolayer
MOKE	magneto-optic Kerr effect
MS	multiple scattering
NW	Nishiyama–Wassermann
PhD	photoelectron diffraction
RFS	renormalised forward scattering
RHEED	reflection high–energy electron diffraction
rlv	reciprocal lattice vector
RRR	residual resistance ratio

SCLS	surface core–level shift
SEMO	surface–enhanced magnetic order
SK	Stranski–Krastinov
SM	simultaneous multilayer
SODS	surface–order dependent state
SPLEED	spin–polarised low–energy electron diffraction
SPPD	spin–polarised photoelectron diffraction
SPSEES	spin–polarised secondary electron emission spectroscopy
SPUPS	spin–polarised ultraviolet photoelectron spectroscopy
SSC	single–scattering cluster
SSE	solid state electrotransport
STM	scanning tunnelling microscopy
STS	scanning tunnelling spectroscopy
SXRD	surface x-ray diffraction
TEM	transmission electron microscopy
TLEED	tensor low–energy electron diffraction
uhv	ultra–high vacuum
UPS	ultraviolet photoelectron spectroscopy
VW	Volmer–Weber
XPD	x-ray photoelectron diffraction
XPS	x-ray photoelectron spectroscopy

CHAPTER 1

INTRODUCTION TO THE RARE EARTHS

Ten billion years ago the rare–earth elements that we see on Earth today were born in a supernova explosion. They account for one hundredth of one percent of the mass of all the elements found in the Earth's crust. Although this may not sound much, bear in mind that this is many orders of magnitude greater than the sum of all of the precious metals. Throughout the aeons of the Earth's history, the rare earths have stayed together with unswerving loyalty. Despite billions of years of being subjected to the physical extremes of geological processes — repeated melting and resolidifying, mountain formation, erosion and immersion in sea water — the rare earths have not separated out into elemental minerals. This is a strong testament to the similarity of many of their physical and chemical properties.

Rock formations resulting from various geological processes can become enriched or depleted in some of the rare–earth elements, and analysis of their relative abundances can yield to geophysicists valuable information on the development of geological formations. It is also generally accepted that the relative abundances of the rare–earth elements in chondritic meteorites represent their overall abundance in the universe, and so this particular branch of rare–earth science is important for an understanding of the genesis of the chemical elements. Compared to such global–scale terrestrial and extra–terrestrial studies, the study of the atomic structure of the surfaces of the rare–earth metals may seem at first to be rather prosaic. It is our aim to show that this is definitely not the case, and that this research is prerequisite for an understanding of many of the properties of the rare–earth metals that are being investigated at present and will be exploited in the future.

1.1 What's in a Name?

The name *rare earth* has its origins in the history of the discovery of these elements. They are never found as free metals in the Earth's crust and pure minerals of individual rare earths do not exist. They are found as oxides which have proved to be particularly difficult to separate from each other, especially to 18th and 19th Century chemists. The early Greeks defined *earths* as materials that could not be changed further by sources of heat, and these oxides seemed to fit that definition. The *rare* part of their name refers to the difficulty in obtaining the pure elements, and not to their relative abundances in the Earth's crust; all of the rare–earth elements are actually more abundant than silver, and some are more abundant than lead. This also explains why the names of some of the rare–earth elements sound similar to each other — what was originally thought to be the earth of a single element was often found subsequently to be a mixture of two or more earths, requiring the hasty invention of more names derived from the original.

In this chapter we present some general background information about the rare earths that help to place the rare–earth metals in their proper context. This includes a brief account of the two–hundred–year history of their discovery, the realisation of their relationship to the other elements in the periodic table, and the uses to which they have been put.

1.2 Discovery of the Rare Earths

There are large deposits of rare earths in Scandinavia, South Africa, China and Australia, but the geographic distribution of 18th Century scientists was such that the Scandinavian deposits were the first to bear fruit. In 1794 the Finnish chemist Johan Gadolin, while investigating a rare mineral found near the town of Ytterby in Sweden, discovered a new earth. He gave it the name ytterbia, which was later shortened to yttria. (The name of Ytterby has been used in various guises to name four of the rare–earth elements, adding to the general confusion of their identities.) The mineral, later named gadolinite, yielded another earth in 1803 through the joint work of Jöns Berzelius and Wilhelm Hisinger and independently by Martin Klaproth. This earth was named ceria, after the asteroid Ceres that had been discovered only two years earlier. Since yttria and ceria had been discovered in a rare mineral, and they resembled closely other earths that were known at that time, they were referred to as rare earths. It was some years later that the English chemist Sir Humphry Davy demonstrated that earths were not elements, but compounds of metallic elements with oxygen. The elements of which yttria and ceria were the oxides were given the names yttrium and cerium. Four decades later it was shown by the Swedish chemist Carl Mosander (who was a student of Berzelius) from his work over the years 1839–43 that the earths then identified as yttria and ceria were in fact oxides of mixtures of elements. He reported that if solutions of the earths were subjected to a long series of fractional precipitations as various salts, then two new elemental substances could be separated from the main component of each of the oxides. The two new oxides found in yttria were named erbia and terbia (the names derived from Ytterby again), and those found in ceria were named lanthana and didymia (from the Greek meaning 'concealed' and 'twin'). Mosander was also the first to extract the rare–earth metals from their oxides, though only in a rather impure form.

The existence of these six new elements and their associated oxides were perplexing to many scientists. The elements appeared to belong to a group that was different from any known at the time. Each of the elements formed the same type of compounds with very similar properties, and the elements themselves could only be distinguished from each other by relatively small differences in the solubilities and molecular weights of the compounds that they formed. In the 1840s and 1850s there was considerable confusion and controversy over the results of further fractionation of the rare earths, with the names of the earths discovered varying from one laboratory to another. Indeed, the level of confusion over the naming of the rare earths was such that in 1860 it was decided, by general agreement of the scientists of the day, to interchange the names of erbia and terbia. From 1859 the use of the spectroscope resolved much of the confusion by providing an objective means to identify the constituent elements. The pattern of lines in the light emission or absorption spectrum of a substance were found to be characteristic of the elements present, and so the products of various fractionation processes could be identified without the ambiguity that characterised earlier attempts.

For much of the next one hundred years fractionation of the rare earths was investigated across Europe and North America. Didymia was shown to be a mixture of the oxides of samarium (1879), praseodymium (1885), neodymium (1885) and europium (1896). Terbia and erbia were resolved into oxides of ytterbium (1878), holmium (1878), thulium (1879), dysprosium (1886) and lutetium (1907). The name lutetium (from the Latin name for Paris) was accepted by all scientists except those in Germany, who referred to it as cassiopeium until the 1950s, when the names were agreed internationally. Gadolinium was discovered, but not named, by Jean-Charles-Galinard de Marignac in 1880. He separated gadolinia from the mineral samarskite (not, as one might have guessed, gadolinite) and some years later Paul-Émile Lecoq de Boisbaudran produced a more pure form of the same earth. With Marignac's approval, he named it after the mineral gadolinite.

Over this period, many dozens of claims were made for the discovery of new rare–earth elements. Most of these were the result of impurities, often from transition metals, causing changes in the apparent molecular weight and emission spectra of their compounds. Without an understanding of the electronic structure of atoms, it was not possible to predict how many rare–earth elements there should be and hence refutation of claims for discoveries of new elements was not a simple matter. The development of the periodic classification of the elements in the late 1800s and early 1900s helped clarify the situation significantly.

1.2.1 The Periodic Table

When Dmitry Mendeleyev first proposed the periodic table of the elements in 1869 he left a blank at the location now occupied by scandium. He realised that an element yet to be discovered, which he called ekaboron, belonged at that location and in 1871 he predicted some of the properties that it would possess. The discovery of scandia in gadolinite by Lars Nilson in 1879, followed by the realisation by Per Cleve that scandium was the 'missing' element ekaboron, provided strong evidence of the validity of Mendeleyev's periodic table and contributed to its acceptance by the scientific community. Although scandium and yttrium now fitted into the periodic table, the placement of the remaining rare–earth elements was proving to be more problematic, as the table structure proposed by Mendeleyev could not accommodate them in any logical manner. Resolution of this puzzle would have to wait until an understanding of atomic structure had developed in the early 1900s.

The English physicist Henry Moseley studied elemental x–ray emission spectra over the years 1913–14 and discovered a relationship between the x–ray frequencies and the atomic number of the element. This relationship made it possible to assign an unambiguous atomic number to any element and thus to verify its location in the periodic table. Moseley concluded that there are 92 elements from hydrogen to uranium and that there are 14 lanthanides with atomic numbers covering the range Z = 58–71.

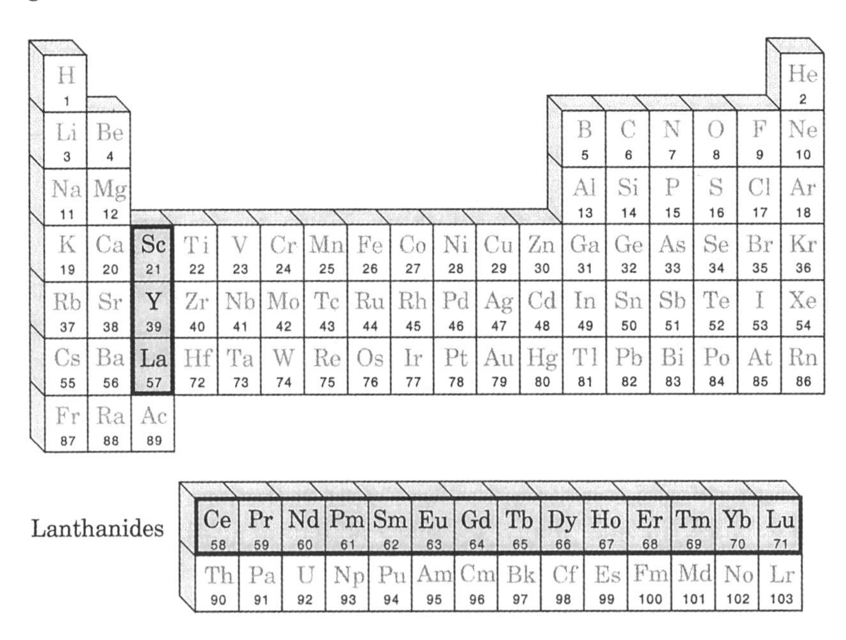

Fig. 1.1 The periodic table of the elements.

At that time, all of the rare–earth elements were known with the exception of element 61. No stable isotopes of this element exist in nature and so it is found only as a by–product of nuclear reactions. Early investigators who mistakenly thought that they had found element 61 prematurely named it illinium and florentium, but it was not until 1947 that one of its radioactive isotopes was unambiguously identified in the fission products of a nuclear reactor. The element was then named promethium, after the Greek Titan who stole fire from the gods.

1.3 Defining the Rare Earths

The rare–earth metals are, by definition, the Group IIIb elements Sc, Y, La and the 14 lanthanides Ce–Lu. The term 'rare earth' has often been applied in the more restricted sense as a synonym for the lanthanides, thus excluding Sc, Y and La. It is reasonable to consider the lanthanides as a group separate from the rest of the periodic table only if the principal interest in these elements is those properties that depend on the occupancy of the $4f$ electron shell, from 0 (La) to 14 (Lu). An important example can be found in the magnetic properties of the lanthanides — the complex, often exotic, magnetic structures observed in alloys and compounds containing these elements are intimately dependent on the lanthanide $4f$ electrons and are thus absent from Sc, Y and La. However, essential though the $4f$ electrons are in giving the lanthanides their character, the grouping together of the 17 rare–earth metals reminds us that they have analogous electronic configurations — the outer three electrons are $(3d\,4s)^3$ for Sc, $(4d\,5s)^3$ for Y and $(5d\,6s)^3$ for La and the trivalent lanthanides. As chemical interactions between atoms are dominated by their outer–electron configurations, this similarity is directly responsible for the rare–earth elements being difficult to separate from each other. Thus, the origin of the 'rare' part of the name lies not in the low abundance (indeed, many are more abundant in the Earth's crust than Pb), but in the difficulty with which any of the elements could be obtained in a pure state by chemical extraction.

Thus, in many respects, Sc and Y can be treated as 'prototype' lanthanides, sharing all of the properties of the lanthanides that are not dependent upon the $4f$ electrons. Playing devil's advocate, we can use the argument in the previous paragraph to extend the definition of 'rare–earth' to include all of the Group IIIb elements, *i.e.*, to include Ac and the actinides as well. Although this does have some logic behind it, including the actinides has two distinct disadvantages. Firstly, the total number of elements in the group reaches 32 (including the 14 lanthanides and 14 actinides), a rather unwieldy number to treat as a group. Secondly, it masks the fact that the character and properties of the lanthanide and actinide elements are quite different due to the differences in the spatial extents (and hence overlaps) of the $4f$ and $5f$ electron wavefunctions. In the lanthanides, the highly localised $4f$ electrons in one atom have essentially zero overlap with those of a neighbouring atom.

However, the same cannot be said of the $5f$ electrons in the actinides due to the greater spatial extent of their wavefunctions. In this respect, the actinides bear more resemblance to the d-block transition metals than they do to the lanthanides. Thus, treating the rare–earth elements as a group distinct from the rest of the Periodic Table is a reasonable practice.

1.4 Applications of the Rare Earths

Although scientific research is the principal destination for the rare earths that are subjected to the highest levels of purification, it is the industrial uses of rare–earth compounds or mixtures of the rare–earth metals that account for the bulk of the rare earths processed throughout the world.

1.4.1 Oxides

One of the principal industrial uses of the rare earths, involving millions of tons of raw material each year, is in the production of catalysts for the cracking of crude petroleum. They also catalyse various other organic reactions, including the hydrogenation of ketones to form secondary alcohols, the hydrogenation of olefins to form alkanes, the dehydrogenation of alcohols and butanes, and the formation of polyesters. The catalytic behaviour of the rare earths can be quite significant, but they are used only in situations where there is no alternative, and less expensive, material that will perform with comparable efficiency.

The rare–earth oxides have been used in various other applications. One of the most widespread uses, responsible for most of the production of ultrapure rare earths, is in the television industry. In the triad of red–green–blue phosphor dots that make up television and computer monitor screens, a mixture of europium and yttrium oxides provides a brilliant–red phosphor. Another large user, which employs rare earths in a number of different applications, is the glass industry. Cerium oxide is more efficient than rouge for polishing glass in the production of lenses for spectacles, cameras and binoculars. Glasses having a combination of low–dispersion and high–refractive index, of use in high quality lens components, can be created using lanthanum oxide. Doping glasses with rare earths can modify their absorption characteristics, making them suitable for use in welders' and glassblowers' goggles. In addition, various compounds of the rare earths have found uses in high–power lasers (for cutting and welding), solid state microwave devices (for radar and communications systems), gas mantles, various electronic and optical devices, and in the ceramic, photographic and textile industries.

1.4.2 Metals

Since the early 1900s the primary commercial form of mixed rare–earth metals has been misch metal, an alloy comprising 50% cerium, 25% lanthanum, 15% neodymium and 10% other rare–earth metals and iron. This metal has been used heavily by the metallurgy industry to improve the strength, malleability, corrosion and oxidation resistance and creep resistance of various alloys, especially steels and other iron–based alloys. In the manufacture of cathode ray tubes, misch metal acts as a getter to remove oxygen from the evacuated tubes. Other elemental rare earths are used as alloy additives — in particular, praseodymium is used in high–strength, low–creep magnesium alloys for jet engine parts. Precision castings of aluminium and magnesium have also been reported to be improved by the addition of other rare earths. The discovery by Auer von Welsbach that an alloy of cerium and iron emits sparks when struck started the flint industry, leading to the refinement of thousands of tons each year of the cerium–laden mineral monazite.

Considerable excitement in the scientific community was aroused by the discovery in the late 1980s of high–temperature superconductivity in compounds of rare earths (in particular, yttrium and lanthanum), copper and combinations of other transition metals. The common link between the various compounds discovered to have high critical temperatures was that they all adopted a crystal structure closely related to the oxide mineral perovskite. The initial fervour pushed up the price of the rare–earth metals yttrium and lanthanum in anticipation of the years ahead when there would be a huge increase in demand for the manufacture of large quantities of room–temperature superconductors. However, this situation was short–lived as subsequent years of intense study revealed that the superconducting properties of the compounds were determined by the behaviour of the atoms in the copper–oxygen planes, and that the role of the rare–earth atoms was secondary.

A significant industrial application of the rare–earth metals is in the production of strong permanent magnets for use in a wide range of electromechanical devices. An often–quoted example is the ultra-lightweight headphones that accompany personal cassette and compact disc players, as these are consumer products with which many individuals will have come into contact in everyday life. This is a rather trite example of a technology capable of producing very powerful and yet lightweight electric motors. When mixed with 'traditional' magnetic metals from the transition metal series, such as iron, cobalt or nickel, some rare–earth metals produce hybrid magnets with exceptional properties that cannot be matched by either rare–earth metals or transition metals alone. In particular, neodymium and samarium have been found to produce materials with highly desirable magnetic properties. The replacement of transition metals with rare–earth metals in permanent magnet materials has clear scientific and technological benefits in the manipulation of the magnetic characteristics of the materials, but the most powerful driving force behind these developments has been the reduction in costs associated with reducing the cobalt content of the materials, due to its high expense relative to other metals.

1.4.3 Elements

Although as yet no link has been made between the atomic structure of surfaces and the radioactive state of the nuclei within the atoms (and it seems unlikely that any will be found), for the sake of completeness we note the use of radioactive isotopes of the rare earth elements in a variety of disciplines — irradiated thulium produces x-rays that are used in portable units for medics, or archaeologists investigating metallic artefacts; yttrium is used in cancer therapy; the relative abundance of lutetium isotopes is used to date meteorites; and the soft β-radiation of promethium is converted to electricity in miniature batteries formed by sandwiches of promethium and silicon.

1.5 The Rare–Earth Metals

Belonging to a common group, it is not surprising to find that many of the rare–earth metals have similar properties. However, the similarities in their chemical properties that are a result of their atomic structure, and hence a part of their definition, are not necessarily reflected in other properties. Indeed, where differences between the individual metals do exist, they can be quite striking. In the following sections some of the basic properties of the rare–earth metals are presented and discussed briefly.

1.5.1 Electronic Configurations

As an introduction to the properties of the rare–earth metals, Table 1.1 lists their electronic configurations, crystal structures and lattice constants. As the electronic configurations of the elements in the rare–earth series have a significant influence on their crystal structures, and hence their surface structures, it is important to consider the systematics of the filling of the outer electron subshells in the lanthanides. The outer electron configurations for the rare–earth elements Pr–Sm and Tb–Tm are different for the atomic and solid state — the divalent $4f^{n+1}\,6s^2$ atomic

Table 1.1 Structural and electronic properties (from Beaudry and Gschneidner 1978)

Element		Z	A	Electron Config	Radius / pm Ionic	Radius / pm Metallic	Crystal Structure	Lattice Parameters a / pm	c / pm	c / a
Scandium	Sc	21	45	$(3d4s)^3$	78.5	164.1	hcp	330.9	526.8	1.592
Yttrium	Y	39	89	$(4d5s)^3$	88.0	180.1	hcp	364.8	573.2	1.571
Lanthanum	La	57	139	$4f^0\,(5d6s)^3$	106.1	187.9	dhcp	377.4	1217.1	3.225
Cerium	Ce	58	140	$4f^1\,(5d6s)^3$	103.4	182.5	fcc	516.1	—	—
Praseodymium	Pr	59	141	$4f^2\,(5d6s)^3$	101.3	182.8	dhcp	367.2	1183.3	3.222
Neodymium	Nd	60	144	$4f^3\,(5d6s)^3$	99.5	182.1	dhcp	365.8	1179.7	3.225
Promethium	Pm	61	145	$4f^4\,(5d6s)^3$	97.9	181.1	dhcp	365	1165	3.19
Samarium	Sm	62	150	$4f^5\,(5d6s)^3$	96.4	180.4	rhom	362.9	2620.7	7.222
Europium	Eu	63	152	$4f^7\,(5d6s)^2$	95.0	204.2	bcc	458.3	—	—
Gadolinium	Gd	64	157	$4f^7\,(5d6s)^3$	93.8	180.1	hcp	363.4	578.1	1.591
Terbium	Tb	65	159	$4f^8\,(5d6s)^3$	92.3	178.3	hcp	360.6	569.7	1.580
Dysprosium	Dy	66	163	$4f^9\,(5d6s)^3$	80.8	177.4	hcp	359.2	565.0	1.573
Holmium	Ho	67	165	$4f^{10}(5d6s)^3$	89.4	176.6	hcp	357.8	561.8	1.570
Erbium	Er	68	167	$4f^{11}(5d6s)^3$	88.1	175.7	hcp	355.9	558.5	1.569
Thulium	Tm	69	169	$4f^{12}(5d6s)^3$	86.9	174.6	hcp	353.8	555.4	1.570
Ytterbium	Yb	70	173	$4f^{14}(5d6s)^2$	85.8	193.9	fcc	548.5	—	—
Lutetium	Lu	71	175	$4f^{14}(5d6s)^3$	84.8	173.5	hcp	350.5	554.9	1.583

configurations become trivalent $4f^n\ (5d\,6s)^3$ in the solid state (where n = 2–5 and 8–12, respectively). Ce is a special case, as a precise value for the $4f$ occupancy and the resultant valency can not be determined uniquely. For the other rare–earth elements, the atomic valency is unchanged upon forming a solid. Along with praseodymium and terbium, cerium is different from the other trivalent rare–earths in that it forms compounds in which it is tetravalent; it is the only rare earth that exhibits a +4 oxidation state in solution.

Surface studies of Pm are conspicuous by their absence as the most stable Pm isotope is radioactive with a half–life of 18 years. Unless the elemental metal or one of its alloys or compounds proves to (or is predicted to) exhibit interesting properties that are absent from those of its neighbours Nd and Sm, this situation is not likely to change in the near future.

1.5.2 Crystal Structures

As can be seen from the table above, the 17 rare–earth metals exist in one of five crystal structures. At room temperature, nine exist in the hexagonal close–packed (hcp) structure, four in the double c-axis hcp (dhcp) structure, two in the face–centred cubic (fcc), and one in each of the body–centred cubic (bcc) and rhombic (Sm-type) structures. This distribution changes with temperature and pressure as many of the elements go through a number of structural phase transitions. The changes with elevated temperature are particularly relevant to the growth of single–crystal samples and will be discussed further in Chapter 4. All of the crystal structures, with the exception of bcc, are close–packed (*i.e.*, the number of nearest neighbours, or coordination number, has its maximum possible value of 12). The close–packed structures can be defined by the stacking sequence of the layers of close–packed atoms, as shown in Fig. 1.2. If the three inequivalent sites on a two–dimensional hexagonal lattice are labelled A, B and C, then the hcp structure is defined by a stacking sequence of $ABAB$..., dhcp is $ABAC$..., fcc is ABC... and the eponymous Sm-type is $ABABCBCAC$... Ideal close packing in the hcp structure is characterised by a lattice parameter ratio of $c/a = \sqrt{(8/3)} \approx 1.633$. As can be seen from Table 1.1, the ratios for the hcp rare–earth metals fall into the range 1.57–1.59, indicating that the atoms in the basal (hexagonal) plane are dilated by ~ 3 % with respect to ideal close packing. The corresponding

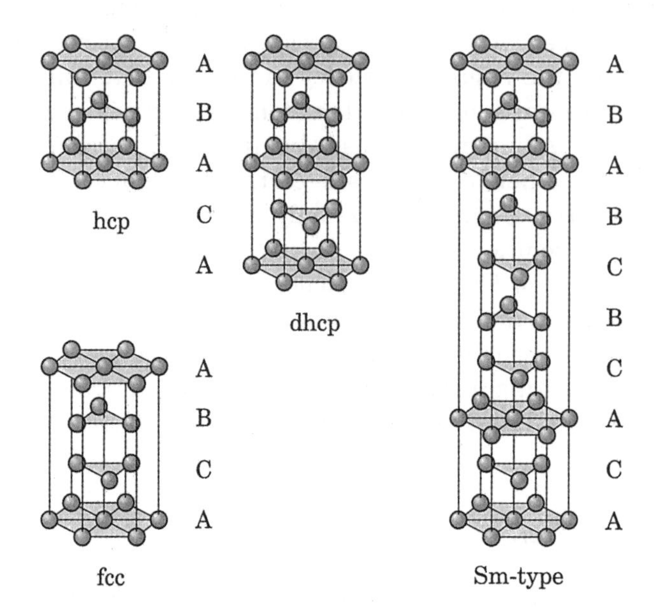

Fig. 1.2 Conventional unit cells for the crystal structures based on the hexagonal lattice.

values for the dhcp and rhombic metals, taking into account the extra atoms in the unit cell, indicate a similar dilation.

If Ce and the divalent metals Eu and Yb are notionally removed from the 17 rare–earth metals, then the remaining 14 can be divided into two major sub-groups: (i) the heavy rare–earth metals Gd–Lu, with the exception of Yb and the addition of Sc and Y, and (ii) the light rare–earth metals La–Sm, with the exception of Ce and Eu. Within each group, the chemical properties of the elements are very similar, with the result that they are almost always found together in mineral deposits (the latter group together with Ce compounds). It is quite common for the terms 'heavy' and 'light' to be applied to the rare–earth elements in the context of these two groups rather than the more usual reference to the atomic number. The adjectives 'yttric' and 'ceric' are also used to describe the heavy and light elements, respectively.

All the heavy rare–earth metals adopt the hexagonal close–packed (hcp) crystal structure and all the light rare–earth metals, with the exception of Sm, adopt the double *c*-axis hcp (dhcp) structure. The rhombic crystal structure of Sm (Sm–type) can be viewed as a rather exotic mixture of one part fcc and two parts hcp (just as the dhcp structure can be viewed as an equal mix of fcc and hcp). Considering only the close–packed crystal structures, there is a systematic variation of the room temperature structures along the lanthanide series as the hcp structure gradually takes over from the fcc:

lanthanide elements	La–Ce		La–Pm		Sm		Gd–Lu
fcc : hcp ratio	1 : 0	→	$^1/_2 : {}^1/_2$	→	$^1/_3 : {}^2/_3$	→	0 : 1
crystal structure	fcc		dhcp		Sm–type		hcp

La and Ce have been included in both the fcc and dhcp categories as they both have a dhcp–fcc transformation close to room temperature. The application of high pressure causes the elements to revert to the previous structure in the series — Gd and Tb adopt the Sm–type structure, Sm becomes dhcp and La, Pr and Nd become fcc. The reason for this behaviour lies with the electronic structures of the rare–earth metals. Skriver (1982) suggested that the sequence of crystal structures could be entirely explained by the variation in d-band occupancy across the series, supporting his argument with bandstructure calculations that showed a systematic decrease in the d-band occupancy for the trivalent lanthanides. Other theories have involved the changing f occupancy across the lanthanide series, but the observation of the pressure–induced hcp → Sm–type → dhcp → fcc sequence for Y (Grosshans *et al* 1982) clearly rules out the direct involvement of f electrons.

All but three of the studies of rare–earth single–crystal surfaces have used samples with crystal structures based on the hexagonal lattice — either bulk hcp or dhcp crystal structures or thin films of close–packed hexagonal layers with an unspecified stacking sequence. The exceptions are the photoemission studies of fcc γ–Ce (Jensen and Wieliczka 1984, Rosina *et al* 1985,1986) which investigated the valence band features of the (100) surface.

As the hcp structure is common to so many studies of rare–earth metal surfaces, it is worth taking a closer look at this crystal structure. It is assumed that the reader is familiar with the Miller index notation used to label crystallographic planes and directions, and in particular the four–index notation that is often used for crystals that have hexagonal symmetry (undergraduate text books on solid state physics, such as

Blakemore 1992 or Kittel 1996, cover the definitions). Note that such familiarity is not essential, as the Miller indices can be treated simply as labels that define the crystal planes and surfaces. As we will be considering only a limited number of planes and directions, in crystals with either hexagonal and cubic symmetry, then the origins of the Miller indices are not strictly required.

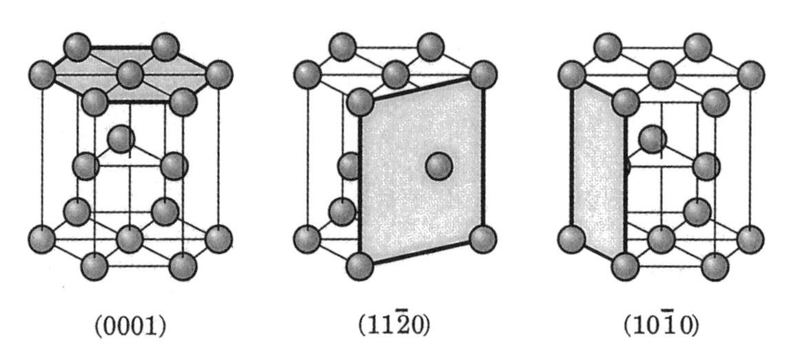

Fig. 1.3 Principal planes of the hcp crystal structure.

The principal planes of the hcp crystal structure and their corresponding Miller indices are shown in Fig. 1.3. Clearly, the $(11\bar{2}0)$ and $(10\bar{1}0)$ planes are less densely packed than the (0001) basal plane (with relative packing densities of 61% and 50%, respectively, assuming ideal close packing). Note that the $(10\bar{1}0)$ plane shown does not cut through any of the atoms in the middle of the hcp unit cell, but the $(11\bar{2}0)$ plane does. The crystallographic structure of surfaces created by cutting the hcp crystal parallel to these planes will be discussed in Chapter 2.

1.5.3 Electronic Structure

On taking a cursory scan through the period table, it may seem that the electronic structure of the lanthanides should be analogous to the transition metal series. In both cases, one of the electronic shells are filling across the series. For the transition metals this is either the $3d$, $4d$ or $5d$ shell, whereas for the lanthanides it is the $4f$ shell. The potential experienced by each electron is determined by the charge densities, and hence the wavefunctions, of all the other electrons in the metal. Clearly, approximations are required to make this a tractable problem, so the many–electron system is treated as a set of soluble one–electron systems.

Although producing an enormous simplification to the mathematics of the solution, most of the important physics is retained. The potentials are fed in to the Schrödinger equation and computers solve the equation numerically to give the wavefunctions of the electrons. These wavefunctions can then be used to recalculate the charge densities and potentials. Thus there is a cycle involving

$$\text{potentials } V \rightarrow \text{ wavefunctions } \Psi \rightarrow \text{ charge densities } \rho \rightarrow \text{ potentials } V$$

that must be self–consistent if the solutions are to have any meaning at all. Using the potentials for a free atom as a reasonable first guess, the cycle can be started and the calculations continued until self–consistency is achieved to within the desired tolerances.

Within such schemes, the electronic structure of extended, or delocalised, valence band states is reasonably well understood, as are the localised core electron states with large binding energies. However, for the lanthanides there is a complication that has impeded calculations of their electronic structures. The $4f$ electrons of the lanthanides have a character that is very different from the s, p and d valence electrons of any other atoms. Spatially, the $4f$ electron wavefunctions bear greater resemblance to core electrons than to valence electrons, and yet energetically they are degenerate with the valence electrons as they have binding energies below ~ 12 eV. This degeneracy of highly localised 'core–like' electrons with the delocalised 'band–like' s and d valence electrons is the problem. Treating the $4f$ electrons as valence electrons in calculations of the electronic structure does not take proper account of their limited spatial extent, and treating them as core electrons does not take account of their interaction with the other valence electrons. Attempts to break this stalemate have been made (see, for instance, Temmerman *et al* 1993) but the rate of progress over the past four decades is a reflection of the difficulty of the problem. The rare–earth metals that have no $4f$ electrons (Sc, Y and La) thus serve as prototypes for the lanthanides, allowing the calculation of electronic structures that can act as benchmarks for all of the rare–earth metals.

Most of the rare–earth metals undergo a transformation of crystal structure at elevated temperatures, which will be discussed further in Chapter 4 in the discussion of crystal growth, but one of the rare–earth metals has a rather unusual transformation that is driven by changes in its electronic structure — the isostructural transformation of Ce between the γ and α phases. Both phases are fcc, but have lattice constants that differ by ~6%. Thus, the phase transformation from γ to α on cooling is manifested as a collapse of the crystal structure. A desire to understand the driving force behind this transformation prompted many theoretical studies of the electronic structure of Ce, and it is now accepted that a change in the relative *d* and *f* occupancy of the valence band is responsible for the crystal transformation (see the reviews by Liu 1978 and Lynch and Weaver 1987, and more recent studies by Gu *et al* 1991, Weschke *et al* 1991, Laubschat *et al* 1992 and Liu *et al* 1992).

1.5.4 Magnetic Structure

Some of the magnetic structures exhibited by the heavy rare–earth metals are shown schematically as a function of temperature in Fig. 1.4. Whilst not presenting the subtlety of the spin alignments in some of the more exotic magnetic structures, it does show some of the diversity of the magnetic phases displayed by some of the metals in the rare–earth series.

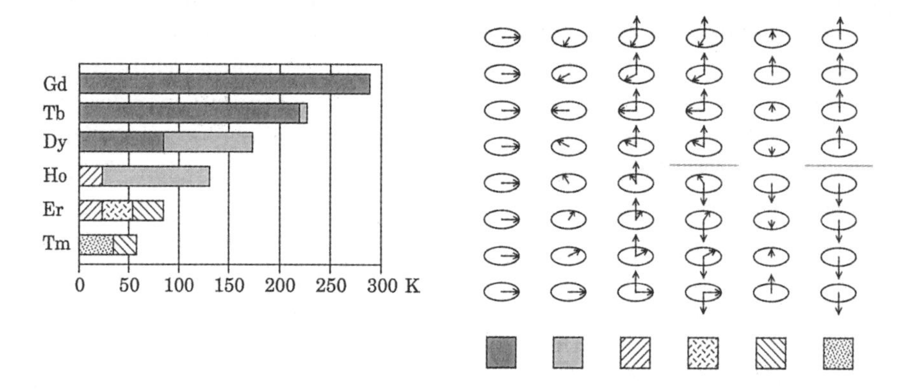

Fig. 1.4 Some of the magnetic phases of the heavy rare–earth metals. For each magnetic phase, the arrows represent the changing magnitudes and directions of the components of magnetic moments relative to the basal planes (circles) from one atomic plane to the next.

The magnetic moments of the rare–earth metals are dominated by the spin contribution from the highly localised 4*f* electrons, and are thus good examples of local–moment ferromagnets. As the 4*f* electron shell can accommodate 14 electrons, a half–filled shell has seven electrons with parallel spins (according to Hunds' rule, the empirical rule in atomic physics that states that in general parallel spins are a lower–energy configuration than anti-parallel spins). Thus, the $4f$ electrons contribute $7\,\mu_B$ to the total magnetic moment of Gd ($\sim 7.6\,\mu_B$), and similarly large contributions to the total moments for the other magnetic rare–earth metals. In contrast to the situation with itinerant ferromagnets (based on the magnetic transition metals), the valence electrons contribute a small fraction of the overall magnetic moment per atom — in the case of Gd, the $5d$ $6s$ valence electrons contribute $0.6\,\mu_B$, less than 10% of the total moment. The magnetic structures of the rare–earth metals and many rare–earth–based compounds are well understood as the result of many decades of experimental study and the development of the local spin–density approximation in calculations of the valence electronic structures of solids.

1.6 Rare–Earth Metal Surfaces

Interest in the surfaces of the rare–earth metals has gone through something of a renaissance over the past ten years, due principally to the discovery of novel magnetic properties of the elemental metal surfaces. Prior to this, surface studies could be categorised as more chemistry than physics.

1.6.1 Early Surface Studies

Over the period 1960–1980 most rare–earth surface studies tended to focus on the chemical properties of the metals and their compounds. These included studies of the adsorption of molecules onto metallic and oxide surfaces, with emphasis on their chemical reactivity and catalytic properties. Many studies involved semiconductor substrates, and in particular Si. The interest here was in the formation of rare–earth silicides at the interface with Si, producing metal–semiconductor contacts with low Schottky barrier heights. Like so many other areas of research related to (or a subset of) rare–earth surface science, the study of rare–earth silicides could justify a book dedicated to that field alone. The scope of this book encompasses the surfaces of rare–earth metals but not those of rare–earth compounds (unless the latter have some bearing on the former) and so studies of rare–earth silicides will not be described here in any detail.

1.6.2 Surface Magnetism

The effect of the presence of a surface on the magnetic properties of metals is currently a very active topic of research. The atomic structure and morphology of a surface influences the behaviour of the valence electrons and hence the magnetic properties of the system. Even in a local–moment system such as a rare–earth–based metal or compound, the behaviour of the valence electrons is crucial as it is these electrons that mediate the exchange interaction between neighbouring magnetic moments localised on the parent atoms. Thus the surface can be thought of as a perturbation to an infinite three–dimensional crystal, which in turn perturbs the electronic structure, which in turn perturbs the magnetic structure. This form of linear thinking is rather restrictive. A more realistic approach is to consider how a redistribution of the electrons near the surface of a material may result in a lowering of the total electronic energy or a change in the magnetic interaction between moments that *drives* the surface to change its structure. The true picture of a surface, if such a concept can be realised, involves a complex interplay between geometric structure (atomic positions and surface morphology) and electronic/ magnetic structure (the behaviour of atomically localised and itinerant electrons). The desire to understand this interplay is the prime mover behind most, if not all, of the studies of rare–earth metal surfaces.

We are not there yet, but the studies described in this book represent some pieces of the jigsaw. More studies will be carried out and more books will be written to provide the remaining pieces. When, or if, we have the complete picture of the relationships and correlations between surface structure and electronic/magnetic properties, then we will have the opportunity to manipulate surfaces to provide materials with the desired properties—surface engineering at an atomic scale.

Further Reading

For those readers who wish to find out more about the physics and chemistry of the rare earths, one of the most comprehensive sources of information is the *Handbook on the Physics and Chemistry of Rare Earths*. Covering all aspects of the rare earths —atomic properties; structural, electronic, magnetic and optical properties of the elemental metals and their compounds, alloys and intermetallics; technological and industrial applications — the handbook now comprises more than 20 volumes.

CHAPTER 2

THE BASICS OF SURFACE STRUCTURE

In this chapter we cover the concepts and definitions that are required for a description of the atomic structure of a surface. Firstly, the distinction is drawn between real surfaces and those rather more idealised surfaces that are assumed to exist. Fundamental to any attempt to describe surface structure is an understanding of bulk and surface crystallography, and so we include a recap of the fundamentals before describing the crystallographic structure of various (ideal) rare–earth surfaces. As the most widespread techniques used to determine surface structure use the diffraction of electrons or photons, the concept and use of the reciprocal lattice in diffraction studies is crucial.

2.1 Real and Ideal Surfaces

A lattice and a crystal, although often confused with each other, are quite distinct. A lattice is an infinite set of (mathematical) points which has translational symmetry and a crystal is the result of placing a set of (physical) objects at each of these points. Implicit in the definition of the crystal is that it is infinite and perfect. In the real world such a material cannot exist but it is accepted practice to use the word in a less strict sense to mean a material that is not a perfect crystal, but has crystalline qualities. Many defects can exist in crystalline materials, such as vacancies, interstitial atoms, dislocations, stacking faults, *etc*, but the existence of surfaces is such a familiar defect of an otherwise infinite crystal that is it often not considered a defect.

The surface of a crystalline material can be defined as the locus of points at which the translational symmetry of the crystal terminates. Many of the solid–vacuum surfaces that exhibit properties of technological interest are not necessarily atomically flat. Indeed, some properties are dependent on a large value for the total surface area of a material, in which case a highly corrugated rough surface would be preferable to a flat surface. It is possible that only a particular surface atomic arrangement is responsible for the desired property, but a rough surface ensures that many different atomic arrangements exist across the surface and so the probability of that particular structure being present is high. Rough surfaces are very difficult to control and characterise, and so identifying that particular surface structure is rather difficult.

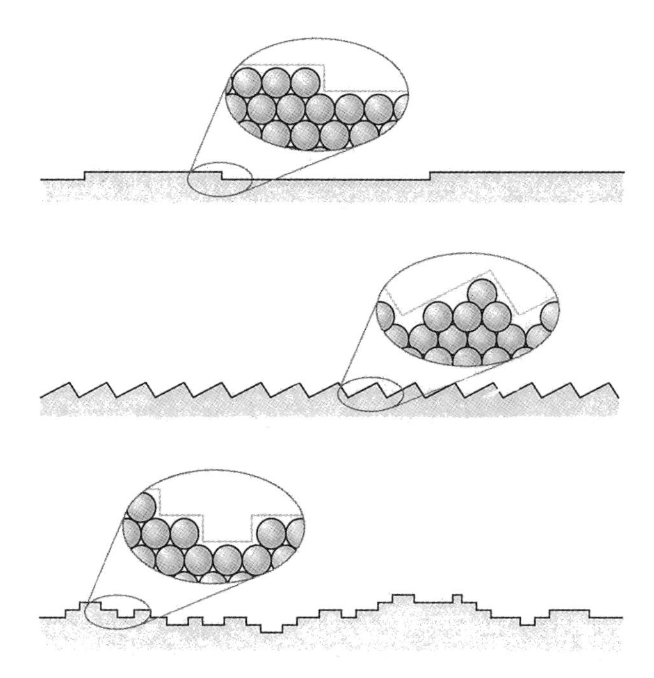

Fig. 2.1 Terraced (top), faceted (middle) and rough (bottom) surfaces.

As a result, the surfaces that are studied experimentally or modelled theoretically, to provide a better understanding of the fundamental physics underpinning those properties, are usually flat. Here, 'flat' is taken to mean that the surface comprises either (i) flat terraces of atoms (large in extent relative to the characteristic length scale of the probe used to study the surface) separated by steps, or (ii) a series of alternating facets of different crystal planes that results in a surface that is flat on a large scale.

Surfaces that are atomically flat over many tens, hundreds or even thousands of nanometres can be created and controlled using a number of different techniques — those that have been applied to the rare–earth metals will be described in Chapter 4. For the remainder of this chapter we will assume that the surface comprises flat terraces that are sufficiently large that they can be considered infinite in extent.

To describe quantitatively the atomic structure of a surface we must understand the principles of surface crystallography. The structures of rare–earth metal overlayers on various substrates have been studied extensively using electron diffraction, and so we must establish the relationship between the real–space lattice, the reciprocal–space lattice and the qualitative features of the resultant diffraction pattern before attempting to describe overlayer structures. A quantitative description of diffraction and its use in surface structure determination will be deferred until Chapter 6.

2.2 Surface Crystallography

The lattices that underlie the crystal structures of all crystalline materials are defined according to the symmetry operations that can be applied to them, bringing the lattices into coincidence with themselves. For instance, a simple cubic lattice has mirror planes parallel to the faces of the cube, three–fold axes of rotation (the body diagonals of the cube) and two– and four–fold axes of rotation (parallel to the edges of the cube). A simple hexagonal lattice has some symmetry operations in common with those of a simple cubic lattice, lacks some that the cube possesses and also has six–fold axes of rotation that the cube does not possess. Listing the complete set of symmetry operations that can be applied to a three–dimensional lattice allows us to define a set of 14 distinct lattices, the Bravais lattices (Blakemore 1992, Kittel 1996). All crystal structures must be based on one of these Bravais lattices.

2.2.1 Two–Dimensional Lattices

A surface is treated as a two–dimensional entity, even though it is clear that we are dealing with a structure that extends a finite distance into the third dimension. The reason for this is that it is the periodicity that determines the symmetry operations that can be applied to the surface. The loss of periodicity in the third dimension has the effect of reducing the number of lattices that are required to describe all possible surface structures, from the 14 Bravais lattices in three dimensions to only five in two dimensions. The lattice vectors that define the periodicity of a lattice are not unique, and neither are the unit cells formed by making a parallelogram from those vectors. A unit cell that has the smallest area is defined as a *primitive* unit cell and examples of these are shown in bold in Fig. 2.2.

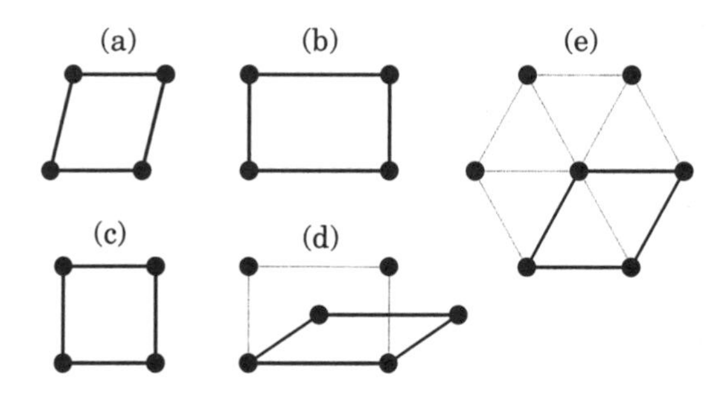

Fig. 2.2 Primitive unit cells of the five two–dimensional Bravais lattices: (a) oblique, (b) rectangular, (c) square, (d) centred rectangular, and (e) hexagonal. The grey lines drawn for (d) and (e) indicate why these lattices have the names that they do.

Note that other unit cells may be chosen in preference to a primitive unit cell, such as the rectangular cell shown in grey in Fig. 2.2(d) for the centred rectangular lattice. This unit cell has twice the area of a primitive unit cell and is often used due to the convenience of having edges that are orthogonal. Such a non-primitive unit cell is defined as a *conventional* unit cell.

By considering all possible ways of adding a collection of atoms, called a *basis*, to these lattices to construct a crystal, it can be shown that there are 17 types of surface structure, distinguished by their different symmetries (the symmetry point groups, space groups and notation used to label them can be found in Woodruff and Delchar (1994)). For some structures, it may not be possible to choose a unit cell that has a single–atom basis, even when the unit cell is primitive.

2.2.2 Definitions of Surface Structure

A few words of definition are required at this point to avoid confusion later. Just as the use of the word 'crystal' has been widened from its strict definition to include a material displaying some crystalline qualities, so the word 'surface' is used by virtually all surface scientists to mean the surface region. The *selvedge* defines that part of a crystal that is influenced by the presence of the true surface. Thus the selvedge extends in two dimensions over the entire surface area of the material, and in the third dimension to a depth that depends on the nature of the material. For some materials, only the outermost atomic layer will be affected by the surface, with the second and third layers being unaware of the existence of the surface above. They would then be considered part of the bulk three–dimensional crystal, or *substrate*. For other materials, deeper layers may be perturbed by the proximity of the surface and so the selvedge may extend for many atomic layers. Atoms of a foreign species (*ie*, a species not present in the substrate) that exist in or above the selvedge layers, through either deliberate or unintentional exposure of the surface to a source of such atoms, are *adsorbates* that form an *overlayer* structure. From this point on, the 'surface' will be taken to mean the surface region comprising the selvedge plus any overlayers.

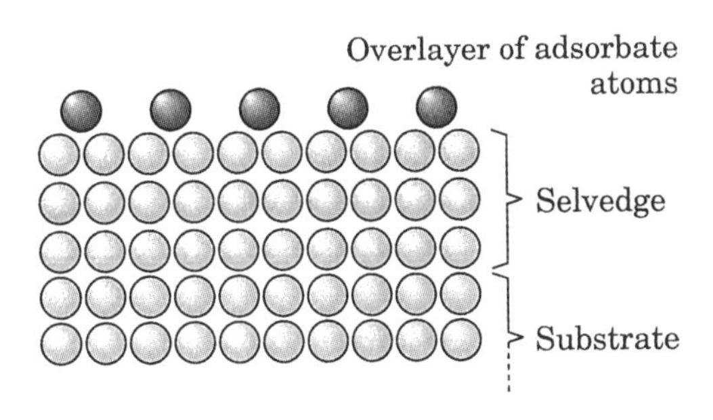

Fig. 2.3 A hypothetical surface illustrating the substrate, selvedge and an overlayer.

The surface of a crystal does not possess the same translational symmetry, or periodicity, as does a three–dimensional crystal — at best, the surface retains translational symmetry in the two dimensions parallel to the surface. If the periodicity (parallel to the surface) of the atoms at the surface is the same as that of the substrate and if the atoms at the surface are separated by a layer spacing that is the same as that of the substrate, then the selvedge is non-existent and the surface is described as *bulk–terminated*. It is more likely that the selvedge layer spacing is different from that of the substrate, in which case the surface is described as *relaxed*. (This has been found to be the case, both experimentally and theoretically, for the surfaces of bulk single–crystal rare–earth samples. Although various studies agree that the relaxation should be non-zero, there is not yet consensus on the magnitudes of the relaxations, or even whether the surface layer spacings should be less than or greater than those of the bulk. This will be discussed further in Chapter 7.) If the periodicity of the atoms in the selvedge is *not* the same as that of the substrate, then the surface is described as *reconstructed*. This change of periodicity at the surface is described by the relationship between the lattice vectors of the surface unit cell (taking into account the structure of the selvedge and any overlayers) and those of the substrate. The notation used to describe this relationship follows one of two conventions, which will be discussed in Section 2.2.4 in the context of overlayer structures.

2.2.3 HCP Surfaces

As all but two of the rare–earth metals adopt the hcp crystal structure, or structures closely related to it, we will take a closer look at the principal surfaces of the hcp structure. The two–dimensional surface unit cells will be described and then the layer–by–layer construction of the hcp and dhcp structures will be examined. The Sm–type crystal structure, having a hcp–like unit cell that is nine atomic layers high, is too complex to consider experimental surface structure determination.

2.2.3.1 Structure Parallel to the Surface

Of the 17 types of surface structure mentioned in Section 2.2.1, three are of particular relevance to studies of the structure of rare–earth metal surfaces. Cutting the hcp crystal structure parallel to its principal planes (as shown in Fig. 1.3) produces the surface structures shown in Fig. 2.4. The surface unit cells (indicated by the parallelograms drawn in bold) are hexagonal for the (0001) surface and rectangular for both the $(11\bar{2}0)$ and $(10\bar{1}0)$ surfaces. It can be seen that the (0001) surface is close–packed and that the $(11\bar{2}0)$ and $(10\bar{1}0)$ surfaces have large unit cells due to their very open structures (Table 2.1). As a consequence of this, the interlayer

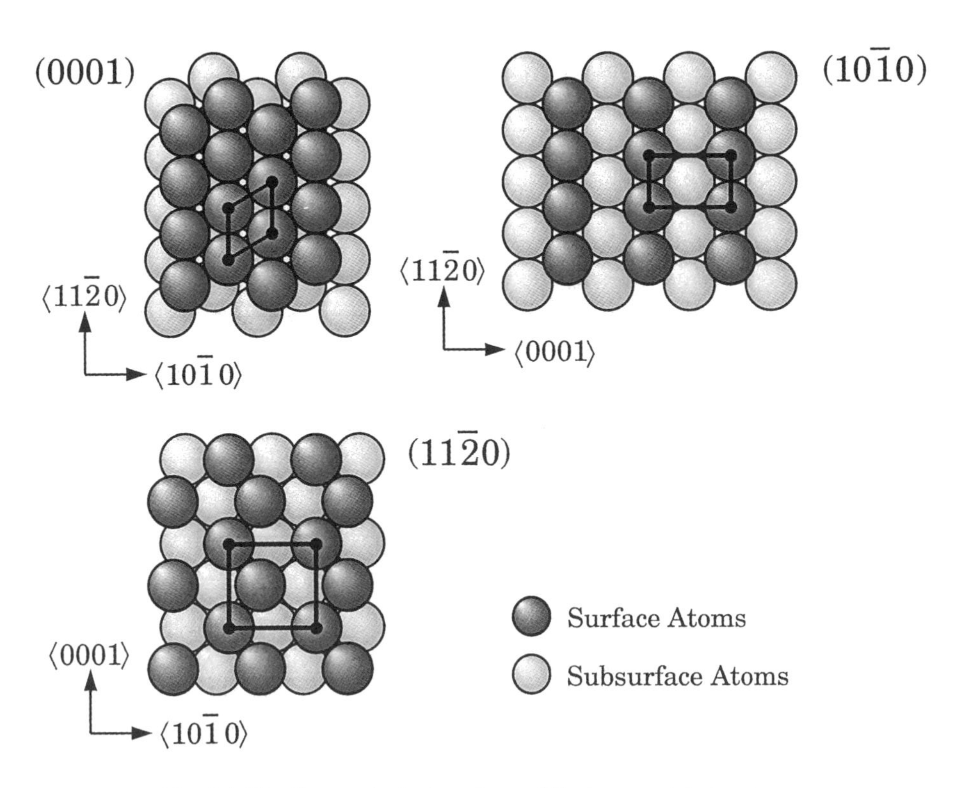

Fig. 2.4 The three principal surfaces of the hcp crystal structure.

Table 2.1 Area packing densities of hcp surfaces, assuming a hard–sphere packing model and ideal close packing.

HCP Surface	Packing Density	Relative to (0001)	Relative to (11$\bar{2}$0)	Relative to (10$\bar{1}$0)
(0001)	0.91	1.00	1.63	1.89
(11$\bar{2}$0)	0.56	0.61	1.00	1.15
(10$\bar{1}$0)	0.48	0.53	0.87	1.00

spacings of these surfaces are relatively small compared to those of the (0001) surface. Note that whereas the (0001) and $(10\bar{1}0)$ surfaces have unit cells with a basis of one atom per unit cell, the $(11\bar{2}0)$ surface structure cannot be represented by a choice of a unit cell with a single–atom basis — the primitive surface unit cell contains two atoms, just as does the primitive unit cell of the hcp structure itself. The consequences of the existence of this second atom in the unit cell will be discussed further in Chapter 7.

The (0001) surfaces of the dhcp and Sm–type crystal structures appear the same as that of the hcp, but the $(11\bar{2}0)$ and $(10\bar{1}0)$ surfaces are rather different. There have been very few studies attempted of these surfaces, presumably because of the problems associated with preparing these surfaces from bulk single crystals and the near–impossibility of growing them as epitaxial thin films (these matters will be discussed in more detail in Chapter 4). The $(11\bar{2}0)$ surface of the dhcp structure is shown in Fig. 2.5 for completeness as this surface of Pr has been studied qualitatively with LEED.

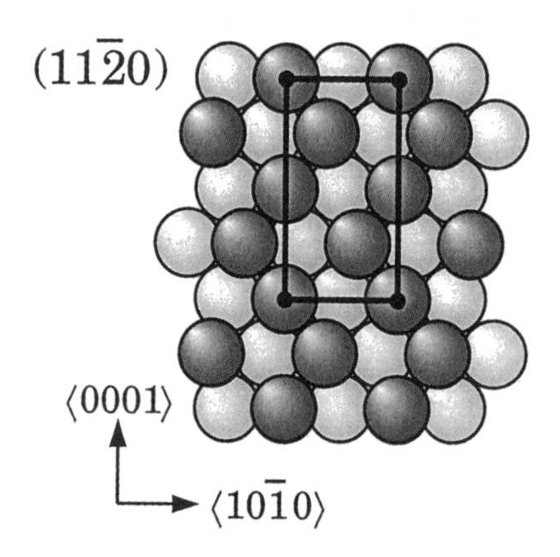

Fig 2.5 The $(11\bar{2}0)$ surface of the dhcp crystal structure.

2.2.3.2 Structure Perpendicular to the Surface

The (0001) and $(10\bar{1}0)$ surfaces are complicated by the fact that they can have one of two different terminations, a point of some significance that will be expanded upon in Chapter 7 when the details of surface structure calculations are discussed. To understand the terminations of the (0001) surfaces of ideal and faulted hcp (and dhcp) crystals we must consider the stacking sequences of these structures. The stacking sequences that define the various close–packed crystal structures are defined by labelling the registries of each layer with respect to the others (Fig. 2.6). Note that the letters used for the layers are not significant in the sense that they can be always be interchanged ($A \rightarrow B, B \rightarrow C, C \rightarrow A$) and so the letter used to label the surface layer is always arbitrary. The

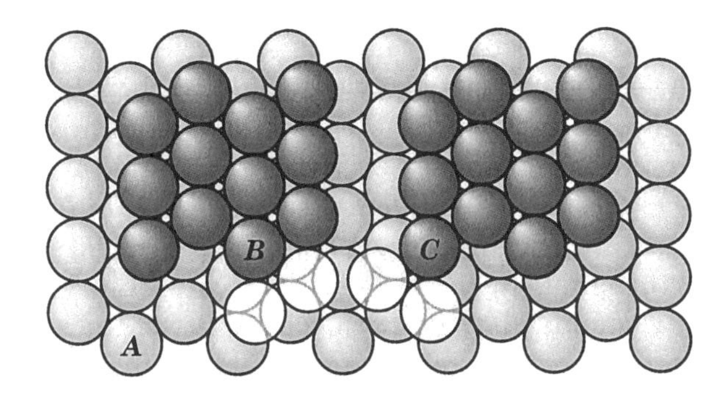

Fig. 2.6 Labelling the inequivalent registries of close–packed layers.

important feature is the repeating pattern of the sequence that results from stacking the layers together to build a three–dimensional close–packed structure. This repeating pattern defines the crystal structure and deviations from the pattern indicate the presence of a stacking fault.

The two terminations of the (0001) surface, conventionally labelled A and B, are equivalent except for a rotation of 180° about the surface normal (Fig. 2.7). As the two terminations have the same surface energy, and are thus totally degenerate, it is expected that the (0001) surfaces of single–crystal samples comprise equal areas of both terminations.

Termination A	Termination B
A	B
B	A
A	B
B	A

Fig. 2.7 The two terminations of the ideal hcp (0001) surface.

The situation for the $(10\bar{1}0)$ surface is somewhat different, as the two terminations of the surface are not related to the stacking sequence, which by definition involves the close–packed layers in the (0001) plane. The two $(10\bar{1}0)$ terminations (again labelled A and B) are characterised by the value of the separation of the surface layer from the first subsurface layer —this value is twice as large for termination A as for termination B. This will be discussed in more detail in Chapter 7.

In addition to differences in the possible terminations of the ideal (0001) surface, it is also possible that this surface has a stacking fault at the surface. The energy balances between the hcp, dhcp, Sm–type and fcc structures can be quite delicate, as evidenced by the pressure–induced crystal transformations mentioned in Section 1.5.2 and the observation of the coexistence of crystallites of different structures within the same boule of rare–earth metal (Shi and Fort 1987). As the surface represents a significant perturbation to the energy balance within the bulk of the metal, due attention must be paid to the possibility of the surface layer or layers having a registry with respect to the underlying layers that is not the same as in the bulk. The simplest example is the possibility of the hcp (0001) surface having a surface layer that has a registry relative to the underlying two layers that is not characteristic of the bulk hcp crystal structure. Fig. 2.8 compares the stacking sequences of three possible terminations of such a surface.

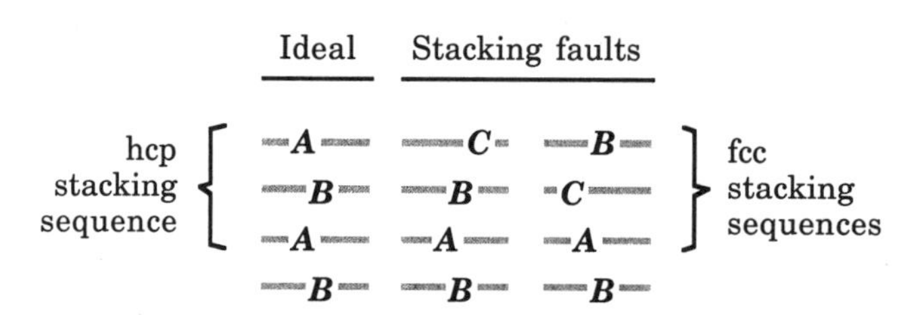

Fig. 2.8 Possible stacking faults at the hcp (0001) surface.

On the left is the ideal, or bulk–terminated, surface. The middle sequence shows a stacking fault at the surface, resulting in the top three surface layers having an *ABC* sequence. On the right is an alternative sequence of *ACB* that has resulted from a stacking fault in the subsurface layer. In general, any layer in a close–packed crystal structure can be described as 'hcp–like' if the layers immediately above and below it have the same registry (as in the bulk hcp structure) and 'fcc–like' if they have different registries (as in the bulk fcc structure). Thus we can describe the surfaces in Fig. 2.8 as having hcp–like or fcc–like terminations.

For the ideal dhcp (0001) surface, the number of possible stacking sequences in the top four surface layers is four and thus the complexity of the calculations involved in surface structure determinations are greater than for hcp. As can be seen from the stacking sequence that defines the dhcp crystal structure (Fig. 1.2), it can be regarded as an equal mix of close–packed layers that are hcp–like and fcc–like (Fig. 2.9). The four

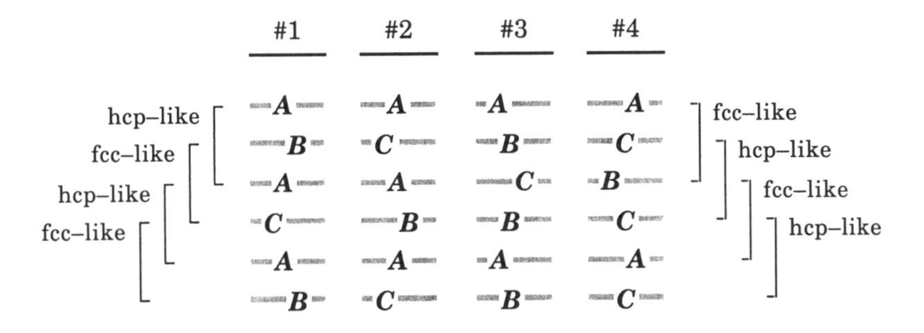

Fig. 2.9 The four terminations of the ideal dhcp (0001) surface.

surface structures can be subdivided into two pairs — the pair that terminate with an hcp–like surface (terminations #1 and #2, using the notation of Dhesi *et al* 1992) and the pair that terminate with an fcc–like surface (terminations #3 and #4). The energy required to introduce a stacking fault into the dhcp structure is less than that required for hcp. As a consequence, it is more probable that we would find stacking faults at the dhcp surface than at the hcp surface. However, translating the surface layer of one of the four ideal dhcp terminations (Fig. 2.9) to an alternative registry produces a structure which, although unique, has the same local structure in the top three layers as one of the other three terminations. It is possible that a stacking fault can occur in the subsurface layers rather than the surface layer (as shown for the right–hand stacking fault structure in Fig. 2.8) but these are much less probable than a surface–layer stacking fault. Most of the techniques used to determine surface structure (to be covered in Chapter 3) are, by their very nature, sensitive to the positions of the atoms in only the top three or four layers. Some can extract structural information from the fourth layer or deeper, but the

Table 2.2 The number of possible hcp and dhcp (0001) surface structures, ideal and faulted.

	Terminations of Ideal Surface	Stacking Fault Permutations Surface Layer Only	Subsurface Layer
hcp	2	4	8
dhcp	4	8	16

accuracy of the structural parameters determined reduces very rapidly with depth. Thus, although the total number of permutations of surface stacking sequence for dhcp can be seen to be quite high (Table 2.2), there are only four that can be distinguished by most surface–sensitive probes.

2.2.4 Notation for Overlayer Structures

As mentioned earlier in Section 2.2.2, when the periodicity of the atoms at the surface differs from that of the substrate, then the change of periodicity is described by the relationship between the lattice vectors of the surface unit cell and those of the substrate. There are two conventions in common use to define these relationships — the matrix notation and the Wood notation.

Defining the primitive translation vectors of the substrate lattice as $\boldsymbol{a}$ and $\boldsymbol{b}$ and those of the surface (selvedge plus overlayer) as $\boldsymbol{a}'$ and $\boldsymbol{b}'$, then the most general and versatile notation for relating these two pairs of vectors is the matrix notation of Park and Madden (1968). If the lattice vectors are related by

$$\boldsymbol{a}' = \mathrm{G}_{11}\boldsymbol{a} + \mathrm{G}_{12}\boldsymbol{b}$$

$$\boldsymbol{b}' = \mathrm{G}_{21}\boldsymbol{a} + \mathrm{G}_{22}\boldsymbol{b}$$

then the surface structure is defined by the four coefficients G_{ij} written in the form of a matrix

$$\mathrm{G} = \begin{bmatrix} \mathrm{G}_{11} & \mathrm{G}_{12} \\ \mathrm{G}_{21} & \mathrm{G}_{22} \end{bmatrix}$$

The type of numbers that form the components of the matrix allow a classification of surface structures:

(i) If all of the components of the matrix are integral then the surface lattice and substrate lattice are defined as being *simply related*. For such a structure, the unit cell of the surface lattice is the same as that of the surface structure as a whole (substrate, selvedge and overlayer).

(ii) If some of the matrix components are rational fractions then the two lattices are commensurate and are defined as being *rationally related*. The unit cell of the surface structure as a whole is determined by the distances over which the two lattices come into coincidence with each other, and hence is larger than those of either surface or substrate lattice. This is the type of surface structure most often produced by the epitaxial growth of rare–earth metals on metal substrates.

(iii) If any of the matrix components are irrational fractions then the surface and substrate lattices are incommensurate, and hence a lattice for the surface structure as a whole cannot be defined.

The notation of Wood (1964) is simpler than the matrix notation, but is consequently less versatile. The Wood notation uses the ratios of the lengths of the primitive lattice vectors for the substrate ($\boldsymbol{a}$ and $\boldsymbol{b}$) and the surface as a whole ($\boldsymbol{a}'$ and $\boldsymbol{b}'$), together with the angle through which one lattice must be rotated to align the two pairs of vectors. If the lattice vectors are related by

$$|\boldsymbol{a}'| = m\,|\boldsymbol{a}| \quad \text{and} \quad |\boldsymbol{b}'| = n\,|\boldsymbol{b}|$$

and the angle between them is ϕ, then the structure of the (hkl) surface is referred to as

$$(hkl)\,(m \times n)\,\mathrm{R}\phi^{\circ}$$

If no rotation of the lattices are required to align their lattice vectors ($\phi = 0^{\circ}$) then the suffix specifying the ϕ angle is dropped.

Implicit in the use of the Wood notation is the assumption that the included angle between $\boldsymbol{a}$ and $\boldsymbol{b}$ is the same as that between $\boldsymbol{a}'$ and $\boldsymbol{b}'$. In the case of the growth of pseudo-close–packed overlayers of rare–earth metals on the (100) surface of tungsten, the overlayers have lattice vectors with an included angle of $\sim 60^{\circ}$ and the substrate has lattice vectors with an included angle of 90°. However, the Wood notation can still be used for this surface structure as it is the lattice vectors of the surface as a whole that are used to generate the indices m and n—the substrate and surface lattices are rationally related and the coincidence lattice has primitive vectors with an included angle of 90°, the same as those of the substrate.

The $(m \times n)$ notation can also be prefixed by the letters 'c' or 'p' to indicate that the cell is centred (in the sense of a centred–rectangular cell) or primitive, respectively. Although a 'centred–square' lattice is not a Bravais lattice (Fig. 2.3), the notation $c(n \times n)$ is still often used in preference to the primitive equivalent as it avoids the need to specify a rotation angle (Fig. 2.10).

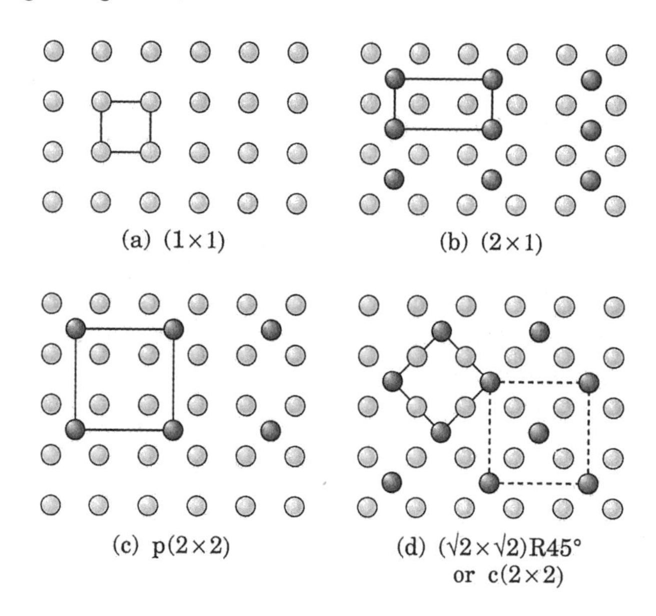

Fig. 2.10 Examples of the Wood notation: (a) a bulk–terminated surface; (b) – (d) overlayer structures. In (d) the primitive (solid) and centred (dashed) unit cells are shown.

2.2.5 *Overlayers on Refractory Metal Substrates*

The refractory metals tungsten, molybdenum and niobium have been found to be most suitable as substrates for the epitaxial growth of thin films of the rare–earth metals. They adopt the body–centred cubic (bcc) structure, which has four atomic planes of interest (Fig. 2.11). Substrates whose surfaces are parallel to these planes have been used for the growth of rare–earth thin films, and the surface crystallography of these are

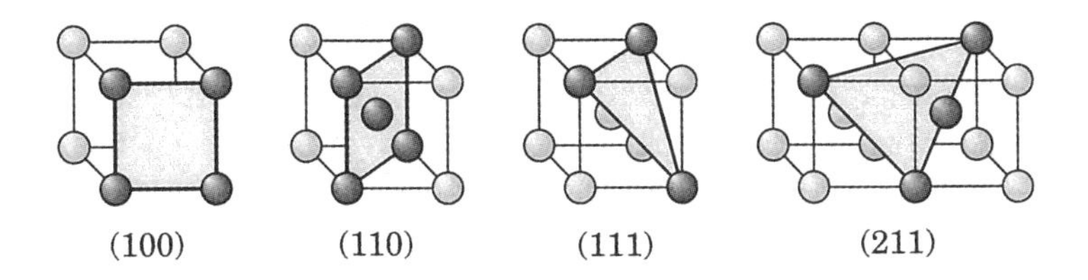

Fig. 2.11 The principal planes of the bcc crystal structure.

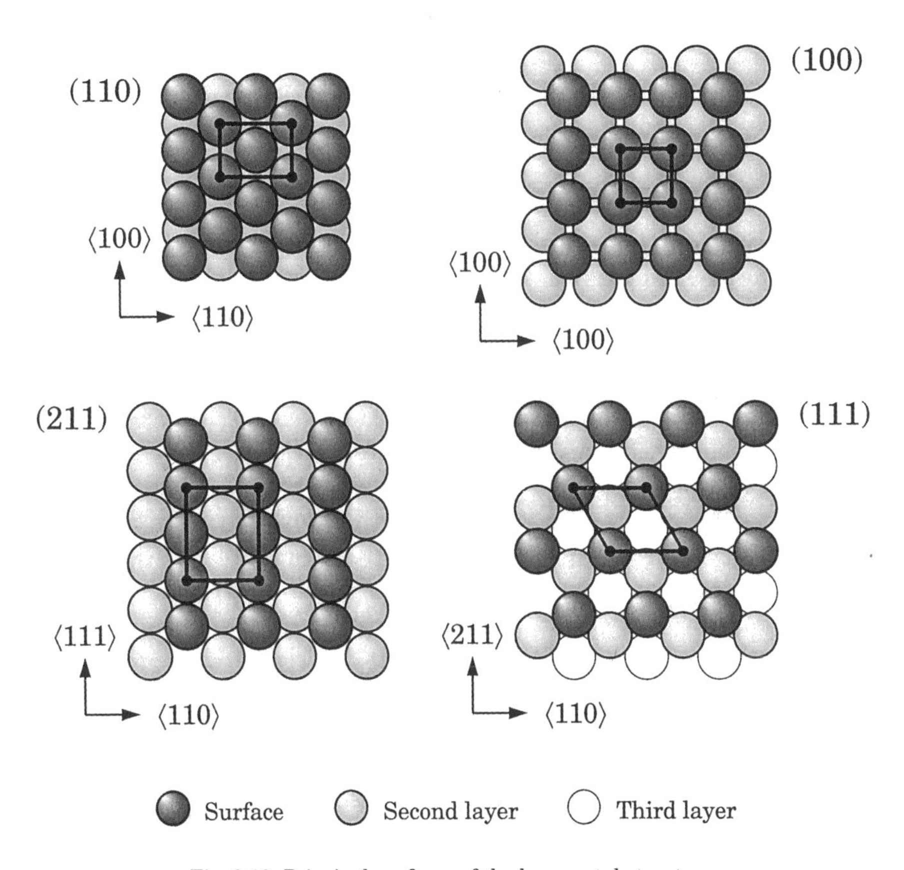

Fig. 2.12 Principal surfaces of the bcc crystal structure.

shown in Fig. 2.12. The relatively high surface energies of the (110) surfaces means that they are not susceptible to adsorbate–induced reconstruction or relaxation, and so can be considered to be rigid foundations on which rare–earth metal films can be constructed. The bcc (110) surface has an area packing density only slightly less than that of the close–packed hcp (0001) and fcc (111) surfaces.

Table 2.3 Area packing densities of bcc surfaces, assuming a hard–sphere packing model.

BCC Surface	Packing Density	Relative to bcc (110)	Relative to hcp (0001)
(110)	0.83	1.00	0.92
(100)	0.59	0.71	0.65
(211)	0.48	0.58	0.53
(111)	0.34	0.41	0.38

Although the (110) surface is the most common substrate for the growth of epitaxial rare–earth metal films, the other surfaces have been used in a number of studies. The (100) surfaces are more open than the (110) surfaces, with a relative packing density of 71%, and hence are more heavily corrugated. The growth of rare–earth metals on the (100) surface is epitaxial, but occurs in two orthogonal domains due to the four–fold rotational symmetry of the (100) surface. The (211) surface has been chosen as a substrate to produce thin films that exhibit considerable anisotropy in the degree of strain and the density of dislocations. The (111) surface has a very open structure, with a relative packing density of only 41%, and might be expected to reconstruct under the influence of the adsorbed atoms. The results of rare–earth metal growth on refractory–metal substrates with each of these surface structures will be discussed in Chapter 5, but before considering the surface crystallography of other metal substrates, it is worth taking a closer look at growth on the bcc (110) surface.

2.2.6 Overlayers on BCC (110) Substrates

Although an understanding of epitaxial metal–on–metal growth has developed since the 1970s (see, for instance, Biberian and Somorjai 1979) interest in the interface structure of an fcc metal with a bcc metal expanded rapidly in the 1980s with the construction of artificial metallic superlattices. Many transition–metal superlattices were created and classified as either *crystalline* or *layered*. In the context of a superlattice comprising alternating layers of metal A and metal B, crystalline means that all layers of A have the same crystallographic orientation as each other and the same is true for all layers of B. A layered superlattice is one in which all layers of A have a common crystallographic axis, but are azimuthally misoriented with respect to each other, and the same is true for all layers of B. A common crystallographic axis means that particular planes are parallel to each other, and it would be expected that these would be the most densely packed planes of the crystal (*ie*, the (0001) planes of hcp, the (111) planes of fcc and the (110) planes of bcc structures). As fcc metals are more common than hcp metals (among the transition metals) most of the crystalline structures found comprised fcc/bcc metal pairs. Thus, much attention was focussed on the growth of fcc (111) metals on bcc (110) substrates and vice versa. The geometric relationships that were determined for the fcc/bcc interfaces were also found to be applicable to rare–earth superlattices. Driven by the novel magnetic properties that could be engineered by sandwiching magnetic rare–earth layers between non-magnetic Y layers, many rare–earth superlattices have been constructed (see, for instance, Majkrzak *et al* 1991). Although some have been made with a growth direction other than [0001] (Du *et al* 1988), the most common growth mode involves layers of close–packed hcp (0001) planes. The preferred substrate for the growth of rare–earth superlattices is sapphire, oriented with a $(11\bar{2}0)$ surface, onto which is grown a Nb(110) buffer layer. This provides a base for an Y(0001) seed layer that is then lattice–matched to the other rare–earth component(s) of the superlattice.

For metals that adopt the fcc or hcp crystal structure in the bulk phase, two preferred orientation relationships exist for the growth of overlayers on bcc (110) surfaces at monolayer and multilayer coverages. They are referred to as the Nishiyama–Wassermann (NW) and Kurdjumov–Sachs (KS) structures and are illustrated in Fig. 2.13. The NW orientation aligns the close–packed $\langle 110\rangle$ rows of the fcc overlayer (or, equivalently, the $\langle 11\bar{2}0\rangle$ rows of the hcp overlayer) to the $\langle 100\rangle$ rows of the bcc substrate. The KS orientation aligns the same overlayer rows to the $\langle 111\rangle$ rows of the bcc substrate. The KS orientation is actually a pair of orientations, as the $\langle 110\rangle$ rows of the fcc overlayer can be aligned to either

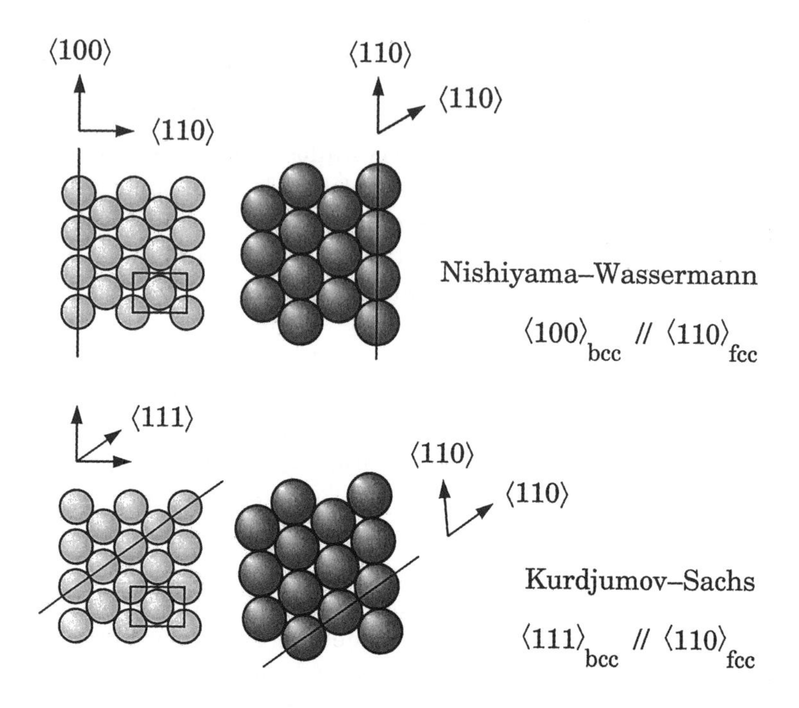

Fig. 2.13 The Nishiyama–Wassermann (NW) and Kurdjumov–Sachs (KS) orientation relationships for a close–packed fcc or hcp overlayer (right) on a bcc (110) surface (left). The Miller indices of the directions in the overlayer are those appropriate for an fcc structure.

of the two row directions in the ⟨111⟩ set of the bcc (110) substrate. As these are equivalent by symmetry, a superlattice that adopts the KS orientation at the interfaces will have a random azimuthal misorientation from one layer to the next–but–one layer, producing a layered, rather than a crystalline, superlattice. The relative orientations of NW and KS differ by

$$\arctan\left(\frac{1}{\sqrt{2}}\right) - \arctan\left(\frac{1}{\sqrt{3}}\right) = 35.3^\circ - 30^\circ \approx 5^\circ$$

and a relatively small variation in the growth conditions (such as the flux of adsorbing atoms, or the temperature of the substrate) can switch the preferred orientation from one to the other.

2.2.7 Overlayers on Other Metal Substrates

A number of non–refractory–metal substrates have been used in the study of rare–earth metal adsorption or thin–film growth, most notably Al, Fe, Ni and Cu, and to a lesser extent also Mg, Co, Ru, Rh, Pd, Ag, Re, Ir and Pt. Discussion of the results of these studies will be dealt with in Chapter 5, but it should be noted that for these systems the substrates cannot be considered to be the inert, rigid foundations that the refractory metals provide. The formation of intermetallic compounds is common, and in some cases is the purpose of the study, and so the resultant structures will be very dependent on the atomic species involved and their temperature treatment. Thus, generalised comments on the structures are of limited validity and care must be exercised before drawing conclusions about the structure of rare–earth metal surfaces grown on these substrates.

2.2.8 Overlayers on Semiconductor Substrates

Interest in the growth of rare–earth metals on semiconductor substrates has been dominated by studies of compound formation at the interfaces of ultra-thin films with the substrates. By far the most common semiconductor substrate used to date is Si, a consequence of the technological importance of the low Schottky barrier heights produced at the interfaces of this element with many rare–earth metals. Due to the reactivity of the rare–earth metals, growth of metallic thin films requires a good lattice match to the rare–earth silicide that inevitably forms at the interface, and thus the comments of the previous section apply here also. This field has been reviewed by Netzer (1995).

2.2.9 Overlayers on Other Substrates

There have been a limited number of studies of rare–earth metal thin film growth on substrates that cannot be categorised as metals or semiconductors. Oxides such as MgO, Al_2O_3, $LaAlO_3$ and Y_2O_3 have been used as some of these materials favour epitaxial growth of metal overlayers (Andersen and Møller 1991a,b, Castro *et al* 1995, Thromat *et al* 1996). Over the past five years, studies of ultra-thin film growth on C_{60} 'substrates', which were usually monolayer films of C_{60} molecules, have concentrated on the effect of the overlayers on the C_{60} molecules rather than on the films themselves (Ohno *et al* 1993, Shikin *et al* 1994, Pan *et al* 1996). The most unconventional of substrates must be polypropylene (Heuberger *et al* 1993,1994), used in a study of the metal–polymer interface in an attempt to understand the adhesion of metal films to polymer surfaces. Again, the metal films were not the principal focus of these studies and thus they are not of particular relevance here.

2.3 Surface Diffraction

The application of diffraction–based techniques to the determination of the atomic structure of surfaces is extensive and so a discussion of the reciprocal lattice and its relationship to diffraction is essential.

2.3.1 The Reciprocal Lattice

The reciprocal lattice can be defined as an abstract mathematical construct that exists not in 'real' space, but in the 'unreal' space referred to as k-space, q-space or reciprocal space. Although a purely mathematical approach will allow us to define the reciprocal lattice and then determine its usefulness in crystallography, we prefer taking a rather more pragmatic approach. The reciprocal lattice can be thought of as a representation of the solutions to the problem of how a wave will diffract from a crystal lattice. Direct visualisation of the reciprocal lattice can best be appreciated through viewing the pattern of spots produced by low–energy electron diffraction (discussed in detail later) from a relatively simple crystalline surface. The symmetry of the diffraction pattern reflects that of the surface crystal lattice, and by changing the energy (and hence the de Broglie wavelength) of the electrons and the angle of the incident electron beam with respect to the surface normal, the effect on the diffraction pattern can be observed in real time. Watching the diffraction pattern expanding and contracting with electron energy, and moving and rotating in space in correspondence with movements of the crystal, helps to establish the concept of the reciprocal lattice as being as much a 'real' entity as the crystal lattice itself. Although such visualisation can be very useful, to use diffraction to determine a real space structure we have to exploit the mathematical relationship between real and reciprocal lattices.

For each real space lattice there is a corresponding reciprocal space lattice that is related to it through a set of mathematical functions. These functions are essentially a representation of the Bragg law expressed for the number of dimensions in which the lattice exists. The diffraction process can be considered from first principles by assuming that a wave is incident upon a scattering medium and produces a scattered outgoing wave. It is assumed that the wavelengths of the incident and scattered waves are the same (which is required for elastic scattering) and that the amplitude of the scattered wave can be calculated only for distances that are large compared to the spatial extent of the scattering medium. Summing the amplitudes of waves scattered from all regions of the scattering medium, taking care to account for the phase differences between the waves, gives an expression for the amplitude of the scattered wave as a function of the wavelength, the scattering strength of the medium and the angle of the outgoing scattered wave relative to that of the incident wave. If the scattering strength of the medium is finite at all points in space, then the scattered wave will, in general, have a finite amplitude for all possible outgoing directions. However, the simplest system of interest for crystalline diffraction has a scattering medium whose strength can be represented by an infinite set of δ-functions. Placing these δ-functions at the positions corresponding to the points of a crystal lattice, it is found that destructive interference between waves scattered from different points in space will produce outgoing scattered waves with zero amplitude, and hence zero intensity, in every direction *except* those specified by a set of equations that relate the directions of the incident and scattered waves to the real space lattice of the crystal. These are called the Laue equations, and they are conventionally written in terms of the scattering vector $\boldsymbol{q}=\boldsymbol{k}'-\boldsymbol{k}$, the difference vector between the scattered and incident wavevectors. The Laue equations are

$$\boldsymbol{q}\cdot\boldsymbol{a}=2\pi h \qquad \boldsymbol{q}\cdot\boldsymbol{b}=2\pi k \qquad \boldsymbol{q}\cdot\boldsymbol{c}=2\pi l \tag{2.1}$$

where $\boldsymbol{a}$, $\boldsymbol{b}$ and $\boldsymbol{c}$ are the crystal lattice vectors and h, k and l are integers. (For a derivation of the Laue equations see, for instance, Blakemore 1992 or Kittel 1996.) Note that for diffraction in two dimensions, only the first two of the Laue equations apply, as will be discussed later. For a given set of vectors $\boldsymbol{a}$, $\boldsymbol{b}$ and $\boldsymbol{c}$, solutions for $\boldsymbol{q}$ in (2.1) can be expressed as

$$\boldsymbol{q}=\boldsymbol{g}_{hkl}\equiv h\boldsymbol{a}^*+k\boldsymbol{b}^*+l\boldsymbol{c}^* \tag{2.2}$$

where $\boldsymbol{a}^*$, $\boldsymbol{b}^*$ and $\boldsymbol{c}^*$ are related to $\boldsymbol{a}$, $\boldsymbol{b}$ and $\boldsymbol{c}$ by

$$\boldsymbol{a}^* = 2\pi\frac{\boldsymbol{b}\times\boldsymbol{c}}{\boldsymbol{a}\cdot\boldsymbol{b}\times\boldsymbol{c}} \qquad \boldsymbol{b}^* = 2\pi\frac{\boldsymbol{c}\times\boldsymbol{a}}{\boldsymbol{a}\cdot\boldsymbol{b}\times\boldsymbol{c}} \qquad \boldsymbol{c}^* = 2\pi\frac{\boldsymbol{a}\times\boldsymbol{b}}{\boldsymbol{a}\cdot\boldsymbol{b}\times\boldsymbol{c}} \tag{2.3}$$

Thus, a solution for scattering exists if the $\boldsymbol{q}$ vector is equal to any one of the vectors $\boldsymbol{g}_{hkl}$ that specifies a point on a lattice in q- or k-space defined by the lattice vectors $\boldsymbol{a}^*$, $\boldsymbol{b}^*$ and $\boldsymbol{c}^*$. These vectors *define* the reciprocal lattice.

Applying the relationship between reciprocal and real lattice vectors defined in equation (2.3), we can determine the reciprocal lattices corresponding to the real space lattices that are of interest to us. For the two–dimensional lattices described in Section 2.2.1, we use the expressions in equation (2.3) in a slightly modified form

$$\boldsymbol{a}^* = 2\pi\frac{\boldsymbol{b}\times\hat{\boldsymbol{n}}}{\boldsymbol{a}\cdot\boldsymbol{b}\times\hat{\boldsymbol{n}}} \qquad \boldsymbol{b}^* = 2\pi\frac{\hat{\boldsymbol{n}}\times\boldsymbol{a}}{\boldsymbol{a}\cdot\boldsymbol{b}\times\hat{\boldsymbol{n}}} \tag{2.4}$$

where $\boldsymbol{a}$ and $\boldsymbol{b}$ are the surface lattice vectors and $\hat{\boldsymbol{n}}$ is a unit vector normal to the surface (conventionally pointing out of the surface — bear in mind that many computer codes used for surface calculations adopt a coordinate system in which the z-axis is normal to the surface, but points *into* the surface). As $\boldsymbol{a}$ and $\boldsymbol{b}$ are both perpendicular to $\hat{\boldsymbol{n}}$, it follows that $\boldsymbol{a}^*$ is perpendicular to $\boldsymbol{b}$ and, similarly, $\boldsymbol{b}^*$ is perpendicular to $\boldsymbol{a}$. Thus, if the unit cell of the real space lattice is rectangular, then $\boldsymbol{a}^*$ ($\boldsymbol{b}^*$) is in the same direction as $\boldsymbol{a}$ ($\boldsymbol{b}$) and so the unit cell of the corresponding reciprocal space lattice is also rectangular. It follows that the reciprocal lattice of a real space square lattice is also square. It can also be shown that the

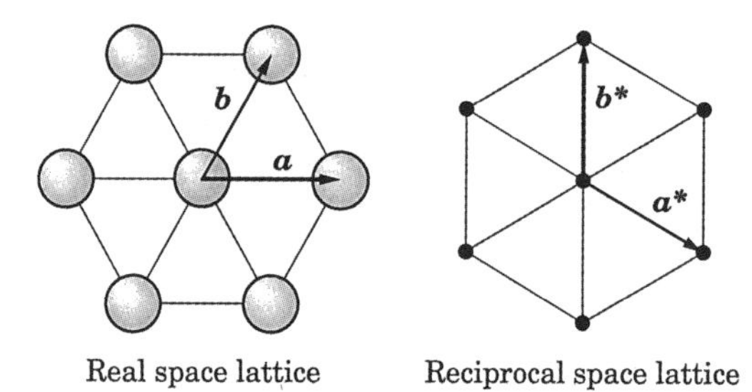

Fig. 2.14 The two–dimensional hexagonal lattice in real and reciprocal space

reciprocal lattice of a real space hexagonal lattice is itself hexagonal (Fig. 2.14), though this is rather less than transparent from a casual inspection of equation (2.4). The absence of orthogonality between the lattice vectors $\boldsymbol{a}$ and $\boldsymbol{b}$ means that $\boldsymbol{a}^*$ ($\boldsymbol{b}^*$) is not in the same direction as $\boldsymbol{a}$ ($\boldsymbol{b}$), and so the resultant lattices in real and reciprocal space are rotated by 90° relative to each other. In three dimensions we use the three expressions in equation (2.3) to deduce that the reciprocal lattice of the simple hexagonal lattice is also simple hexagonal, rotated about the c-axis by 90° as for the two–dimensional lattice. The reciprocal lattice of fcc is bcc, and the reciprocal lattice of bcc is fcc (the symmetry resulting from the fact that the expressions in (2.3) have the same form if the $\boldsymbol{a}$, $\boldsymbol{b}$ and $\boldsymbol{c}$ vectors are expressed in terms of $\boldsymbol{a}^*$, $\boldsymbol{b}^*$ and $\boldsymbol{c}^*$).

2.3.2 The Ewald Construction

The directions in which diffracted waves will emerge from an illuminated sample, assuming that the wavevector of the incident wave and the crystal lattice are known, can now be determined schematically. By drawing a region of the reciprocal lattice and judiciously plotting the incident wavevector, the solutions $\boldsymbol{q} = \boldsymbol{g}_{hkl}$ can be determined from the Ewald construction, an example of which is shown in Fig. 2.15.

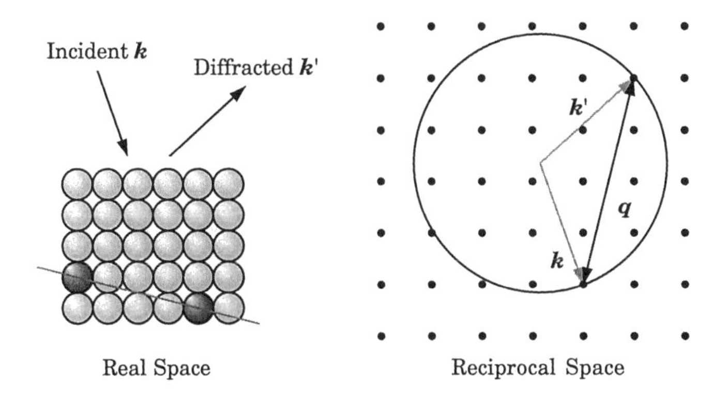

Fig. 2.15 The Ewald construction in reciprocal space (right) for a wave ($\boldsymbol{k}$) incident on a cubic crystal predicts a diffracted wave ($\boldsymbol{k}'$) because the scattering vector $\boldsymbol{q} = \boldsymbol{k}' - \boldsymbol{k}$ is equal to a reciprocal lattice vector. The planes of atoms responsible for the diffracted wave are perpendicular to $\boldsymbol{q}$, and one such plane is indicated (left).

In three dimensions, the Ewald construction comprises (i) a set of reciprocal lattice points, (ii) the incident wavevector $\boldsymbol{k}$ drawn such that it terminates on one of the lattice points, and (iii) a sphere of radius k, centred on the origin of the vector $\boldsymbol{k}$. If the surface of the sphere intersects any other lattice points, then a vector joining the origin of $\boldsymbol{k}$ to this point is the wavevector $\boldsymbol{k}'$ of a diffracted wave, as the $\boldsymbol{q}$ vector must span two lattice points, representing a solution to the Laue equations (or, alternatively, the Bragg law). The same principles apply to diffraction in two dimensions, but the reduced dimensionality of the construction gives rise to significant differences in the solutions.

The effect of the loss of periodicity resulting from the existence of a surface can be considered from a number of viewpoints. A crystal with a surface can be thought of as an entity that has finite repeat distances in the two dimensions parallel to the surface and an infinite repeat distance in the dimension perpendicular to the surface. This results in a reciprocal

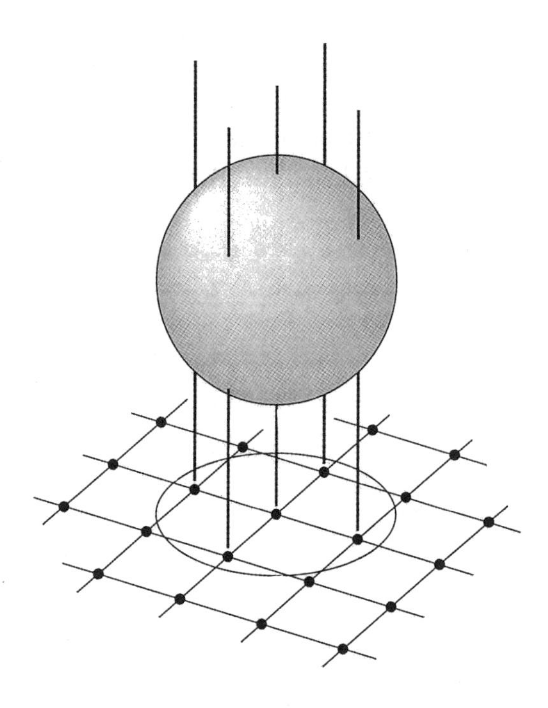

Fig. 2.16 The Ewald construction in two–dimensional reciprocal space. Only those rods that pass through the Ewald sphere are shown for clarity.

lattice that has lattice constants parallel to the surface that are finite, but a 'lattice constant' perpendicular to the surface that is infinitely small. Thus, the rows of lattice points perpendicular to the surface form a line, or rod, in reciprocal space. An alternative approach is to return to the Laue equations (2.1), which assumed an infinite set of scattering points, and consider the effect of the introduction of a surface. Removal of the scattering points from 'above' the surface effectively means that the third of the Laue equations is no longer required to be strictly true, as total destructive interference from the remaining points is no longer complete and so a finite scattering intensity will result for all values of $\boldsymbol{q} \cdot \boldsymbol{c}$. The first two Laue equations remain valid as the set of scattering points is still infinite in the dimensions parallel to the surface. Diffraction from a crystal with a surface thus involves a reciprocal lattice that is not a three–dimensional lattice of points, but a two–dimensional lattice of rods. To visualise the geometry of diffraction from a surface the concept of the Ewald construction in reciprocal space is again used, but now the diffraction condition is met whenever the Ewald sphere intersects a lattice rod (Fig. 2.16).

Although the diffraction can be described as two–dimensional, the Ewald construction still involves a sphere, not a circle, as the incoming wave can be diffracted out of the plane defined by its wavevector and the surface normal. The Ewald sphere and rods make a three–dimensional construction, but looking at two–dimensional slices through the construction can be instructive due to their simplicity. Fig. 2.17 shows vertical slices through the Ewald constructions for a number of possible diffraction geometries. The top diagram of Fig. 2.17 shows what happens to a wave incident normally on a crystal surface. The wavevector has a magnitude slightly larger than the separation of rods in the reciprocal lattice, and as a result there are six solutions to the diffraction condition — the incident wave continues through the crystal, a wave is diffracted directly back from the surface (zero–order diffraction, or reflection), and four other waves propagate either into the crystal or out of the surface (first–order diffraction). The middle diagram of Fig. 2.17 shows the effect of increasing the energy, and hence the magnitude of the wavevector, of the incident wave. Again there are six diffracted waves, but the angles of four of these with respect to the incident wave have changed. This could have been deduced easily enough from a casual inspection of the Bragg law, but the Ewald construction becomes particularly useful when the geometry of

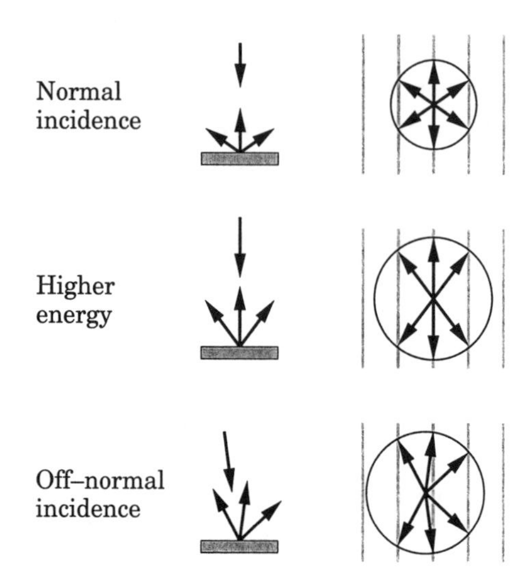

Fig. 2.17 Using the Ewald construction to determine the directions of waves diffracted from the surface of a crystal.

diffraction is less symmetric, such as in the case for off–normal incidence. The lower diagram of Fig. 2.17 shows such a geometry and allows us to visualise the directions of the diffracted waves without recourse to solving equations. Note that if the energy of the wave is low enough such that the wavevector has a magnitude less than a half of the separation of rods in the reciprocal lattice, then only two diffracted waves are present — the incident and reflected wave. Note also that these waves will always exist, regardless of the energy or angle of the incident wave.

So far, it is the relationship between real and reciprocal space lattices that has been discussed. Defining the reciprocal lattice and using the Ewald construction is a convenient method to predict the directions in which waves will be diffracted from a crystalline material, but they tell us nothing about the *intensities* of the diffracted waves. Conversely, the arrangement of atoms in the basis has no affect the directions of diffracted waves (as these are determined by the periodicity) but does influence their intensities due to the interference between waves scattered from the individual atoms. Thus we can summarise diffraction very succinctly: diffraction directions give the lattice, diffraction intensities give the basis.

2.3.3 *Indexing the Lattice Points*

Taking a section parallel to the surface through the reciprocal lattice rods, producing a two–dimensional lattice of points, each point can be indexed by the amount of momentum transfer imparted to the associated diffracted wave (in units of the reciprocal lattice vectors of the substrate). By this convention, the zero–order diffracted (or reflected) wave is indexed as (00) and all the other waves diffracted by the substrate have integer indices. This notation will be used extensively in the context of the quantitative analysis of low–energy electron diffraction, to be discussed in some depth in Chapters 6 and 7.

2.3.4 *Reconstructions and Overlayers*

If the surface is reconstructed or has an overlayer structure, then the periodicity of the surface will be larger than that of the substrate and hence the corresponding reciprocal lattice will be smaller (Fig. 2.18 illustrates this for an overlayer). Note that the additional points that appear

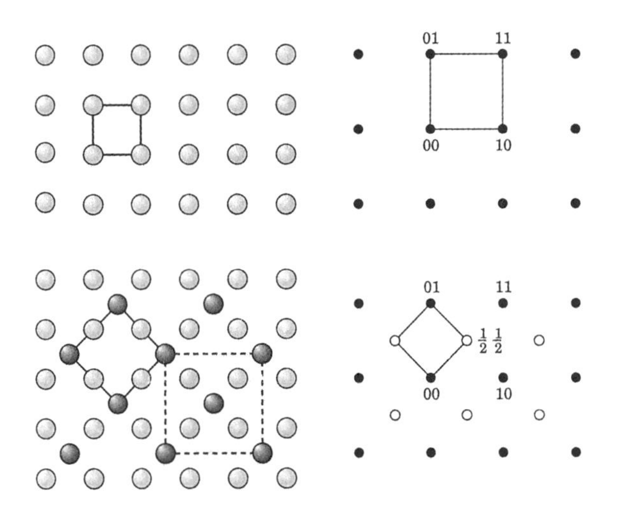

Fig. 2.18 Indexing the reciprocal lattice points of a simple surface (top) and one with an overlayer structure (bottom). The extra reciprocal lattice points produced by the overlayer are shown as open circles. The primitive unit cells of the real and reciprocal lattices are shown as solid squares and the conventional unit cell of the c(2 × 2) overlayer structure is shown as a dashed square.

in reciprocal space have fractional indices. The intensities of the diffracted waves corresponding to these points carry information about the relative positions of the atoms within the basis of the surface unit cell. If the reconstruction or overlayer are commensurate with the substrate, then a subset of the fractional–index points will be coincident with the integer–index substrate points. At such points, waves diffracted by the surface and the substrate will interfere and hence their intensities carry information about the relative positions of the surface and substrate atoms, an observation of some importance when trying to determine the registry of a surface unit cell with respect to that of the substrate.

CHAPTER 3

SURFACE STRUCTURE TECHNIQUES

The principal techniques that have been applied in surface structural studies will be described in this chapter, together with some examples of their application to the rare–earth metals. Although the diffraction–based techniques that will be discussed in Section 3.1 have been essential in surface structure determination, there a number of other techniques that have found supporting roles and two of them are outlined in Sections 3.2 and 3.3. Scanning tunnelling microscopy (STM) is the most direct technique for studying the structure of a surface at the nanometre scale. Atomic–resolution imaging is not always possible on metallic surfaces, and, even when it is, the information provided rarely extends below the top atomic layer. These properties make STM a valuable real–space complement to reciprocal–space diffraction techniques. Photoelectron diffraction (PhD) may at first appear to be in the wrong section, as it may seem more appropriate to discuss it alongside the other diffraction–based techniques in Section 3.1. However, PhD is fundamentally different from these techniques for a number of reasons, and these will be discussed.

3.1 Electron and X-Ray Diffraction

In the previous chapter we have been describing the diffraction of waves from crystals or crystal surfaces without specifying the nature of the incident wave. Although all the principles of diffraction apply equally to beams of x-rays or electrons, the fundamental differences between the interactions of electromagnetic radiation and charged particles with the atoms in a solid lead to quite distinct approaches to both the experimental techniques and the data analysis associated with each type of wave.

3.1.1 Low–Energy Electron Diffraction (LEED)

The diffraction of electrons is intrinsically surface–sensitive if the energy of the electrons is chosen to be ~ 10–500 eV. In this energy range, the electrons experience strong inelastic and elastic scattering, both of which contribute to the surface sensitivity. Strong inelastic scattering means that electrons propagating more than a few atomic layers have a high probability of being inelastically scattered, and hence will not contribute to the intensity of an (elastic) diffracted wave. A short mean–free–path, comparable to a few atomic layer spacings, makes it unlikely that electrons that travel deep within the crystal arrive back at surface unscathed. Strong elastic scattering contributes to the surface sensitivity because layers deep below the surface experience a significantly reduced intensity of incident electrons due to strong backscattering from the layers above.

3.1.1.1 LEED Optics

The construction of a LEED system that can be used for qualitative or quantitative surface structure analysis is shown in Fig. 3.1. As these systems comprise a number of electrostatic elements to accelerate, focus

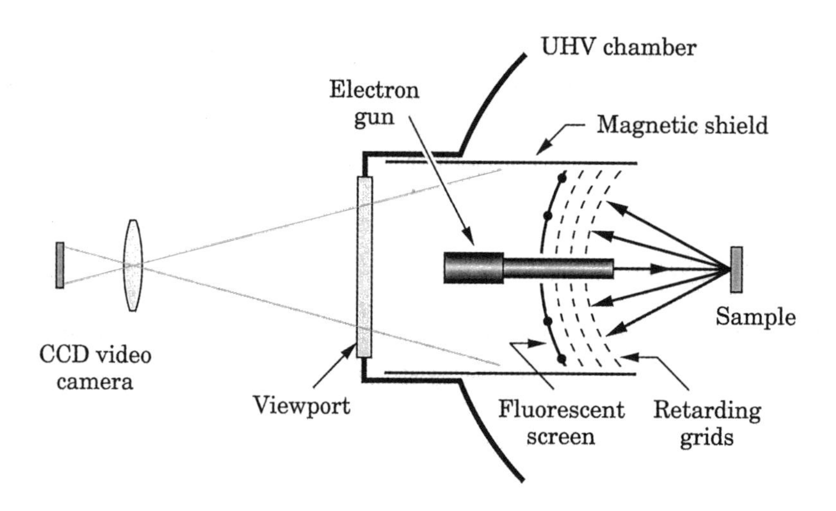

Fig. 3.1 A typical experimental arrangement for LEED optics. The diffraction pattern can be captured and digitised using a video framegrabber and the diffraction spot intensities can be stored in a computer for subsequent analysis.

and retard the electron beams, they are often referred to as LEED optics. An electron gun produces a beam of ~ 1 μA (or rather less in specialist low–current optics designed for surfaces that are sensitive to beam damage) at the desired energy. The sample is usually positioned such that the incident beam is normal to the surface, unless it is the reflected (zero–order) beam that is of interest, and some of the diffracted beams are directed back towards the optics. The electrons in these beams travel in straight lines across the field–free region between the sample and the first of a number of metal spherical–sector grids (both held at earth potential). One or more of the subsequent grids are set to an adjustable retarding potential close to that of the filament in the electron gun. This ensures that electrons that have been inelastically scattered from the sample, losing a few electron volts of energy, cannot pass through the grid. The optimum number of grids in the optics is determined by a compromise between the desire to obtain the highest contrast in the diffraction pattern and the finite transmission of each grid. Increasing the number of grids makes it possible to reduce the 'leakthrough' of electric fields between the grids, but at the expense of reducing the overall intensity of the diffraction pattern. Many of the LEED optics produced by commercial manufacturers use a three–grid design. Once through the retarding grids, the electrons have very little kinetic energy and so they are accelerated by ~ 5 kV onto a fluorescent screen to make the diffraction pattern visible from either behind the sample or, in rear–view LEED optics, from the other side of the phosphor–coated glass screen.

If the sample is placed at the centre of curvature of the spherical grids then the resultant diffraction pattern has a very useful property. As can be seen from the Ewald construction in Fig. 2.17, for a normal–incidence beam the surface–parallel component of the wavevector of a diffracted beam is simply related to the reciprocal lattice of the sample. If the diffracted beams were to be imaged by allowing them to impinge onto a flat fluorescent screen (Fig. 3.2), then the positions of the diffraction spots would present a distorted view of the reciprocal lattice. The spherical geometry of LEED optics not only ensures that the electrons are always travelling perpendicular to the grids, but also that the diffraction pattern viewed is an accurate representation of the two–dimensional reciprocal lattice. Strictly, this is true only if the diffraction pattern is viewed from an infinite distance, and so if the pattern is imaged by photographing the screen or digitising the signal from a video camera, then account must be

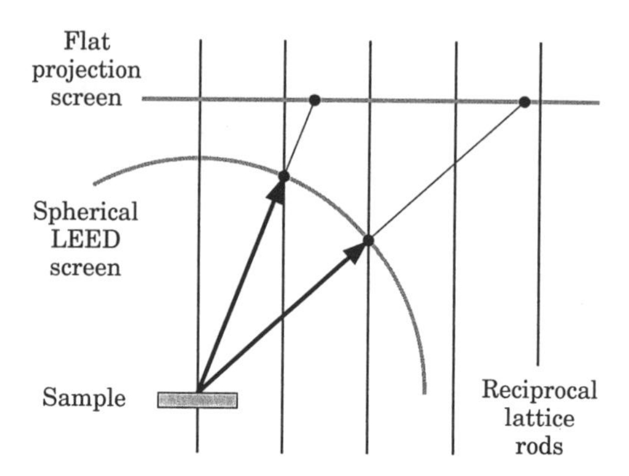

Fig. 3.2 Projecting diffracted LEED beams onto a flat screen distorts the appearance of the reciprocal lattice. Projecting onto a spherical screen produces no such distortion and so the diffraction pattern can be interpreted visually more readily.

taken of the screen curvature. For a known distance between the imaging device and the LEED optics, converting the apparent positions of diffraction spots into the actual directions of the diffracted beams requires simple geometric functions.

3.1.1.2 Measuring Intensities

Qualitative analysis of the diffraction spot positions in a LEED pattern provides direct access to the two–dimensional reciprocal lattice and hence the surface lattice of the sample. However, surface structure determination requires far more than simply the lattice — the positions of the atoms that constitute the basis must be determined. These cannot be determined by casual inspection of the diffraction pattern, as all possible arrangement of atoms within a surface unit cell will produce diffraction patterns with spots in the same positions. For a full surface structure determination the intensities of the diffracted beams must be analysed quantitatively. The intensity of a diffracted beam can be measured by placing a Faraday cup in front of the first grid of the LEED optics and measuring the resulting current. Mechanically simpler alternatives that require no additional apparatus within the uhv chamber have become more popular in most systems. The visual brightness of a spot on a phosphor screen is a function of the intensity of the diffracted beam (although this relationship is not linear as a phosphor screen will saturate at high beam intensities) and so measuring the spot intensity from outside the chamber is a convenient approach. For some systems this has been carried out using spot photometers, but now charge–coupled device (CCD) video cameras and computer framegrabbers are much more common. The video–based techniques have a number of advantages, including the acquisition speed to capture the entire diffraction pattern in a fraction of a second and the ability to store the images for later off–line analysis (Fig. 3.3). These features are particularly relevant for the

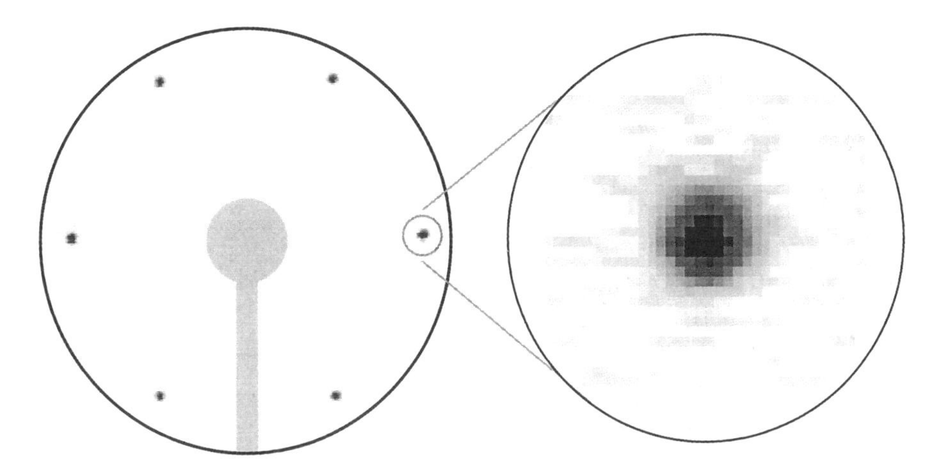

Fig. 3.3 Video image of a LEED pattern (left) from Sc(0001). At a beam energy of 50 eV only the six spots corresponding to first–order diffraction can be seen (the zero–order spot is hidden by the electron gun and the second–order diffracted beams emerge from the surface at angles that are too large to be observed on the screen). Expanding the image around one diffraction spot (right) shows the individual pixels of the image, from which the relative intensities of the diffraction spots can be calculated. Adapted from Barrett *et al* (1993).

analysis of spot intensities from surfaces that are sensitive to contamination from residual gases in the vacuum chamber, as is the case for the rare–earth metals, or for diffraction patterns that are short–lived due to electron beam damage.

The strong scattering of electrons in solids that is responsible for the surface sensitivity of LEED is also responsible for one of the most important restrictions on its use in surface structure determination. Strong scattering implies multiple scattering — it is likely that an electron in a LEED experiment experiences many elastic scattering events before escaping from the surface and contributing to the intensity of a diffracted beam. Measuring the intensity of a diffracted beam is trivial compared to the problems associated with determining what surface structure must exist to produce a diffracted beam with that intensity. Because of the existence of multiple scattering, there is no easy *direct* way to take a set of diffracted beam intensities and 'invert' the data from reciprocal space to produce the desired surface structure in real space. The best that can be done is to calculate the diffracted intensities from a number of model structures that are thought to be close to the unknown surface structure, and to compare these calculated intensities with those measured in a LEED experiment. The closest match then indicates the most likely surface structure, with the confidence level being determined by a goodness–of–fit parameter. In practice, this is done for many different diffracted beams over a wide range of electron energies to discriminate between models which differ only by small displacements of the constituent atoms. Experimentally, this involves measuring the intensities of spots as a function of the voltage used to accelerate the electron in the LEED gun, and so is often referred to as LEED *I–V*. The calculations that account for multiple scattering and their comparison with experimental data will be covered in more detail in Chapter 6.

Although full structure determination requires multiple–scattering calculations, a simple kinematic analysis (*ie*, one that assumes only

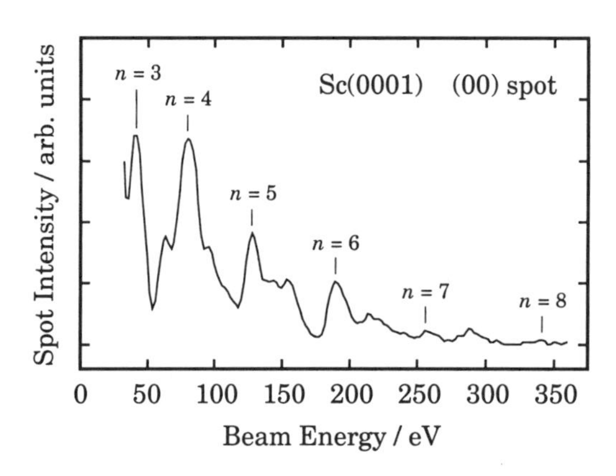

Fig. 3.4 *I–V* spectrum of the (00) LEED spot from Sc(0001). The nth order Bragg peaks produced by constructive interference between the surface and subsurface layers are marked. Adapted from Barrett *et al* (1993).

single–scattering events) allows the identification of some of the features in an I–V spectrum. The peaks in the I–V spectrum of the (00) spot of Sc(0001) shown in Fig. 3.4 can be labelled with the order of Bragg diffraction between the surface and subsurface (0001) layers. Up to the sixth order, the energies of the prominent peaks in the spectrum are predicted from kinematic calculations that assume no surface relaxation. Note that other peaks and shoulders appear in the I–V spectrum that cannot be explained by a kinematic analysis — these can be reproduced theoretically only using full dynamical (multiple–scattering) calculations.

3.1.2 Reflection High–Energy Electron Diffraction

Although LEED is a ubiquitous technique in any surface science laboratory, there is another form of electron diffraction that can give useful information about surface structure. At energies of 10–100 keV electrons have relatively long inelastic mean–free–paths (~ 10 nm) and so to achieve acceptable surface sensitivity requires a specific diffraction geometry. The incident electron beam is brought in at grazing incidence and the diffracted beams at grazing exit angles are observed on a fluorescent screen on the opposite side of the sample from the electron gun. The

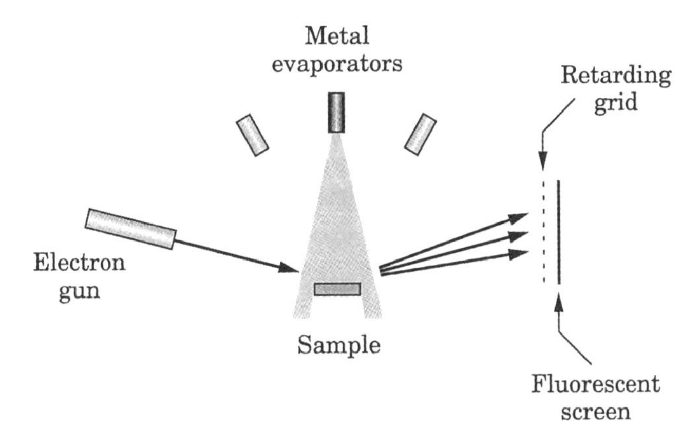

Fig. 3.5 Reflection high–energy electron diffraction (RHEED) experimental geometry.

complex optics associated with LEED are not required for RHEED as the higher electron energies means that suppression of inelastically scattered electrons is not so important and there is no need for additional acceleration of the electrons to produce fluorescence.

By having the incident and diffracted electron beams at grazing angles to the surface, in contrast to the normal–incidence and off–normal diffracted beams in a LEED experiment, the region directly in front of the sample is left unobstructed. This is one of the principal advantages of this technique over LEED, as it allows thin film growth to proceed from evaporators placed close to the sample normal. RHEED is used extensively in molecular beam epitaxy (MBE) systems as a real–time monitor of semiconductor crystal growth, but it has also been used to a lesser extent in the study of metal–on–metal growth.

In reciprocal space, the high energy of the electrons means that the Ewald sphere is large compared to the separation of the reciprocal lattice

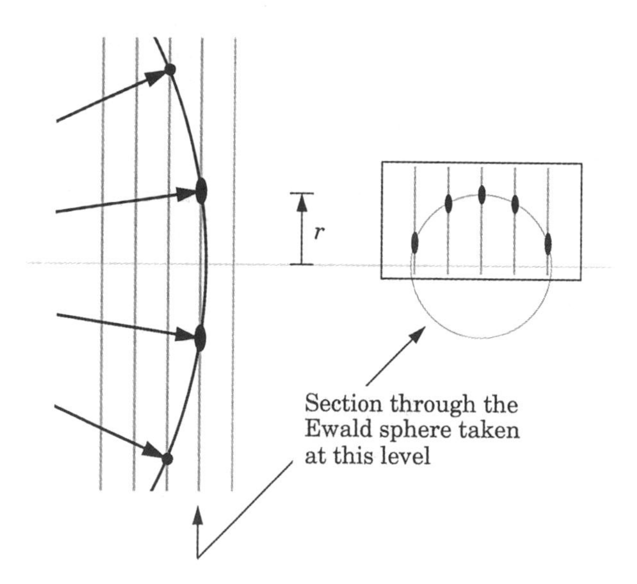

Fig. 3.6 Part of the Ewald construction for a RHEED experiment. A vertical section through the centre of the Ewald sphere (left) has been drawn for an electron energy of ~ 10 keV. The RHEED spots on the fluorescent screen (right) are observed to lie on a circle whose radius r depends on the incident and exit angles.

rods. As can be seen from Fig. 3.6, the grazing geometry has the consequence that rods of finite width (due to imperfect surface order) will give rise to diffraction beams that are elongated in the direction normal to the surface. Thus, all RHEED patterns show spots that are streaked,

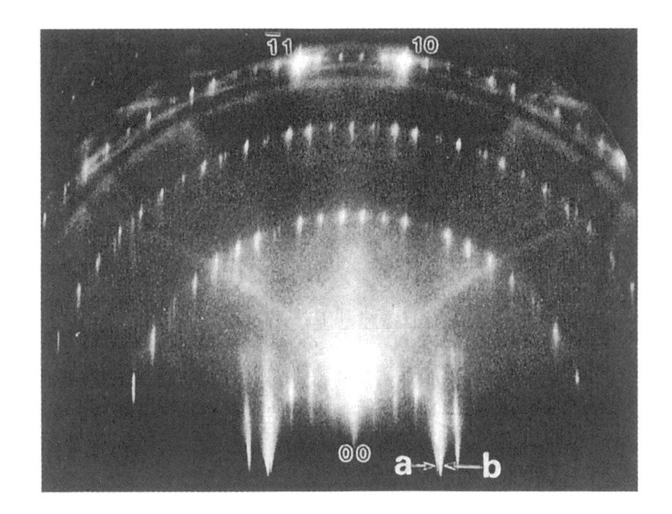

Fig. 3.7 RHEED pattern from 1.5 ML of Ce grown on Mo(110). The labels 'a' and 'b' correspond to first–order streaks of the first and second Ce monolayer, respectively. From Tanaka *et al* (1995b).

with the extent of streaking dependent on the angle of intersection between the Ewald sphere and the reciprocal lattice rod. Figure 3.7 shows a RHEED pattern from a thin film of Ce grown on Mo(110) taken from Tanaka *et al* (1995b). Comparison of the positions of the spots corresponding to the Mo substrate and those of the first and subsequent Ce layers allows the structural parameters and the growth mode of the Ce film to be determined.

Some confusion over the interpretation of RHEED patterns resulted from the fact that a rough surface can produce sharper diffraction spots than a well–ordered surface. Although at first this seems rather counter–intuitive, it can be explained by transmission electron diffraction through protrusions on the surface that are a few atomic layers high. The pattern of diffraction spots produced by transmission will be characteristic of the three–dimensional atomic structure of the sample and thus will lack the streaking that results from the rods associated with surface diffraction. However, disorder parallel to the surface will increase the width of the reciprocal lattice rods and hence increase the streaking. Thus, the reduction of surface order can lead to a RHEED spots becoming either more streaky *or* apparently sharper, depending on the nature and the spatial scale of the disorder. Confusion was inevitable.

Although quantitative comparison with theory has not been developed in the same way as for LEED, the sensitivity of RHEED to changes in surface morphology make it well–suited to the study of MBE growth processes. By comparison, LEED will self–select the regions of the surface that are well–ordered and can be relatively insensitive to large areas of the sample surface having little or no crystallographic order.

Further Reading

The characterisation of ultra-thin films based on the analysis of RHEED data using dynamical diffraction theory has been presented for the Dy/Si system by Mitura *et al* (1996).

3.1.3 Surface X-Ray Diffraction

In contrast to the case for electrons, the diffraction of x-rays is not intrinsically surface sensitive. X-rays of wavelengths comparable to typical crystal lattice constants penetrate into solids to large depths ($\sim \mu$m) because the scattering cross–sections are $\sim 10^6$ smaller than those for low–energy electrons. If this were not the case, then x-ray diffraction would not have established itself as the principal technique for bulk crystal structure determination. However, it is precisely this weak interaction between x-rays and solids that gives x-ray diffraction one important advantage over electron diffraction — multiple scattering events are much less probable and so data analysis should be much more straightforward. Thus if x-rays are to be used in surface structure determination, then methods must be found to enhance the surface sensitivity.

One approach is to limit the penetration of the incident x-rays from the surface by choosing a grazing–incidence geometry. If the grazing angle is below the critical angle for the material, typically a few mrad, then total external reflection can be achieved. This occurs because the refractive index of x-rays in solids is less than, but very close to, unity. Penetration of the x-rays into the surface is then limited to an evanescent wave that has a decay length of a few nm. The grazing–incidence geometry effectively reduces the cross–sectional area of the sample as seen by the incident beam, such that a few millimetres of sample is foreshortened to a few microns. Focussing a beam of x-rays into a spot of this size, with a flux sufficiently high to enable a surface diffraction signal to be measured with reasonable counting statistics, is a non-trivial feat. The development of synchrotron radiation sources, generating high fluxes of well–collimated x-rays of the required energy, has enabled surface x-ray diffraction (SXRD) to evolve over the past two decades.

Even with the limited penetration afforded by a grazing–incidence geometry, x-rays diffracted from the top surface layers still coexist with those produced by subsurface diffraction which may be characteristic of the bulk crystal structure. The surface sensitivity can be improved further providing that the surface lattice is different from that of the substrate. Measuring the intensities of x-rays produced by fractional–order diffraction (corresponding to points in reciprocal space that are *not* coincident with the substrate reciprocal lattice, as discussed in Section 2.3.4) improves the surface sensitivity significantly.

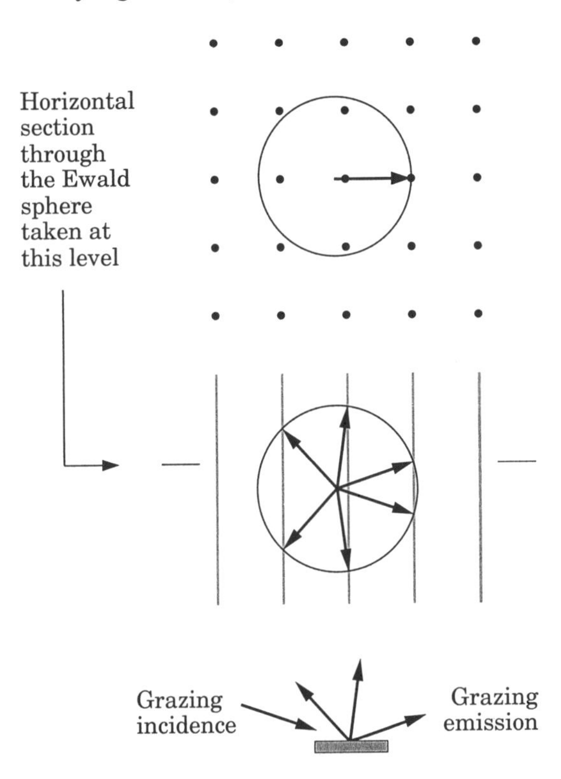

Fig. 3.8 Ewald construction for grazing–incidence surface x-ray diffraction. The grazing angle has been drawn as greater than typical critical angles (~ mrad) for clarity.

Just as in the case for LEED, surface structure determination using x-ray diffraction requires intensity measurements for as many diffracted beams as it is possible to acquire, but meeting this requirement experimentally is significantly more complicated due to the scattering geometry. The Ewald construction shown in Fig. 2.16 is appropriate for a typical LEED experiment, but for surface x-ray diffraction the construction shown in Fig. 3.8 is more representative of typical scattering geometries. Here, the magnitude of the wavevector of the incident wave is comparable to that of a low–energy electron, but the requirement of grazing incidence means that there is one grazing emission (zero–order diffraction, or reflected) beam and a number of other beams diffracted through large angles. To measure the intensities of grazing emission beams corresponding to different reciprocal lattice rods it is necessary to change the azimuthal angle of the incident beam. Fig. 3.9 shows horizontal sections through the Ewald spheres corresponding to two different azimuthal angles. In both cases, the grey circles and wavevectors correspond to the geometry of Fig. 3.8 as a reference.

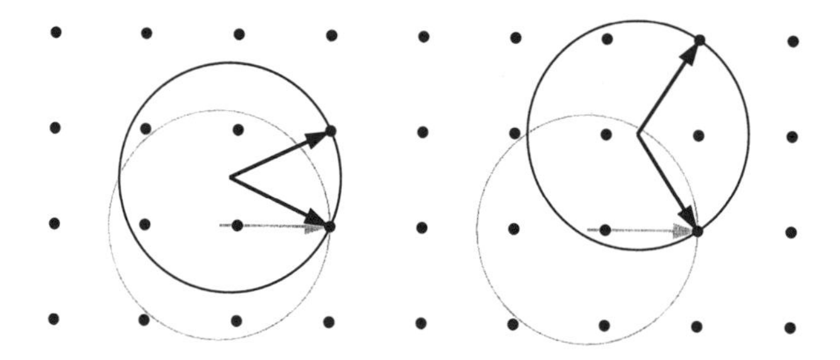

Fig. 3.9 Grazing incidence and emission diffraction for different azimuthal angles of ~ 25° (left) and ~ 55° (right) with respect to the original incident beam.

In all experimental systems the source of the x-ray beam is fixed and so the sample and x-ray detector must be rotated to change the scattering geometry. Bearing in mind that this must be done such that the x-ray beam remains in a grazing–incidence geometry to an accuracy of ~ mrad, it can be seen that this places severe demands on the production of sample surfaces that are sufficiently flat and also on the accuracy and reproducibility of sample manipulation. The latter can be achieved, as for bulk diffraction experiments, using large diffractometers — a massive mounting supports multiple circles that move relative to one another to allow four, five or six degrees of freedom to orient the sample with respect to the incident beam and detector. The extra complexity suffered by surface x-ray diffraction experiments is the need to maintain the uhv environment around the sample at all times, and thus the diffractometer must be coupled to a uhv chamber that has the usual collection of surface preparation and characterisation techniques (such as LEED and Auger electron spectroscopy). As the equipment associated with most surface science chambers is heavy (and often delicate) it is preferable to keep the chamber fixed and allow the diffractometer to move the sample, with restricted movement on some of the axes of motion, through a flexible coupling. Stainless steel or mu–metal uhv chambers are effectively opaque to x-rays, and so an SXRD chamber must be fitted with x-ray transparent 'windows' of a light metal such as Be to let the x-rays in and out of the chamber without significant attenuation. Compared to a LEED experiment, for which selecting a different diffraction beam entails setting a suitable energy for the incident electron beam and deciding which of the spots on the LEED screen corresponds to the selected beam, an SXRD experiment can be seen to be much more involved. However, the simplified analysis of the diffracted beam intensities due to the lack of multiple scattering makes the investment worthwhile.

So far we have looked at how SXRD can be used to study the two–dimensional structure of the surface. Measuring the intensities of x-rays produced by fractional–order diffraction, although desirable to enhance the surface sensitivity, has a significant limitation — no information regarding the registry of the surface unit cell relative to that of the substrate can be extracted from these diffracted beams. To determine this registry, the intensities of diffracted beams that have contributions from both the surface and substrate must be measured. Diffracted beam intensities at points corresponding to the three–dimensional reciprocal lattice of the substrate are not suitable, as the contribution from the substrate will be many orders of magnitude greater than that of the surface and so interference between them will be insignificant. To circumvent this problem we consider the effect of the truncation of the substrate lattice by the surface.

3.1.3.1 Crystal Truncation Rods

The diffraction from an infinite three–dimensional crystal can be described by a reciprocal lattice of δ-functions. The truncation of the crystal at a surface (ignoring any complications due to surface relaxation, reconstruction or overlayers) has the effect of spreading the δ-functions perpendicular to the surface, creating crystal truncation rods (CTRs). The intensity of diffracted beams measured along these rods will vary, with maxima at positions corresponding to the bulk Bragg peaks and (non-zero) minima in between. Our model of diffraction from a surface and substrate now involves reciprocal lattice rods produced by the two–dimensional surface plus CTRs produced by the truncation of the three–dimensional substrate.

Returning to the problem of determining the registry of the surface unit cell relative to that of the substrate, we can now see a solution. Diffracted beam intensities can be measured along surface reciprocal

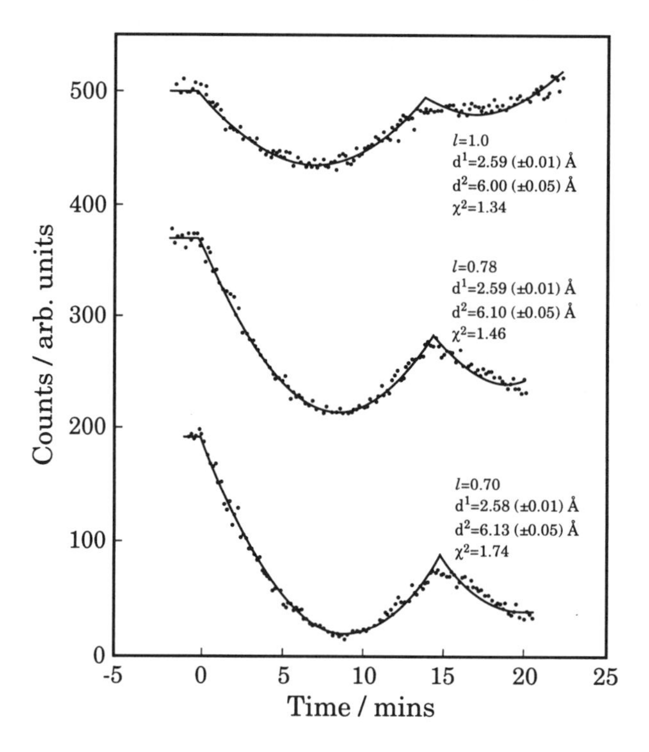

Fig. 3.10 SXRD data for the growth of Sm on Mo(110). The intensities of the $(00l)$ CTR at three values of l are shown as a function of time during the growth. Fitting the data using a theoretical model of film growth (solid lines) provides structural information such as the surface relaxation (d_1) and the overlayer packing density. Adapted from Nicklin *et al* (1996).

lattice rods that are coincident with CTRs. The substrate contribution will dominate close to the CTR maxima, but at the CTR minima the contributions from the surface and substrate will be comparable, and thus interference between them will produce measurable effects.

An example of how the intensities of x-rays along CTRs can be used to determine structural parameters of thin films and their growth characteristics is shown in Fig. 3.10. By fitting the data from the early stages of growth to a two–level model of layer–by–layer (*ie*, FM) growth the perpendicular layer heights (d_1 and d_2) and the layer densities can be determined (Nicklin *et al* 1996). The sharp cusp predicted by the theory for FM growth is not reproduced in the data, indicating either island formation or the incorporation of second layer atoms into the first layer (Stenborg and Bauer 1987b).

Further Reading

The applications of SXRD to semiconductor and metal surfaces and interfaces have been reviewed by Robinson and Tweet (1992).

3.2 Scanning Tunnelling Microscopy

The first practical demonstration of the STM by Binnig *et al* (1982) marked the start of a revolution in the study of surfaces. However, a decade passed before the first STM image of a rare–earth metal surface was published (Dhesi *et al* 1995).

The principal features of the STM are shown in Fig. 3.11. A sharp metal tip (usually W or a PtIr alloy) is moved close to a conducting surface by one or more piezo–electric crystals. These can take the form of a tripod of three crystals aligned along orthogonal axes, a tube, or sometimes more elaborate constructions, although only the former is illustrated here for simplicity. Motion of the tip is achieved by application of voltages to the piezo–electric crystals, causing them to expand or contract according to the polarity of the voltage applied. By fixing the ends of the crystals to a rigid frame, the tip can be moved freely inside a cube whose dimensions may be many microns. In most STMs, macroscopic movement (~ mm) of the tip relative to the sample is achieved by movement of the sample using a piezo–electric crystal with coarser control. A bias voltage is applied to sample, from a few tens of millivolts up to a few volts, and the resultant current that flows between tip and sample is measured. Modelling the tip

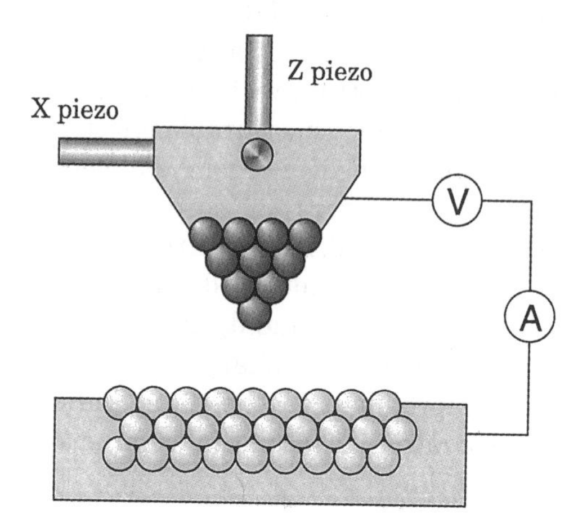

Fig. 3.11 Principal features of the scanning tunnelling microscope

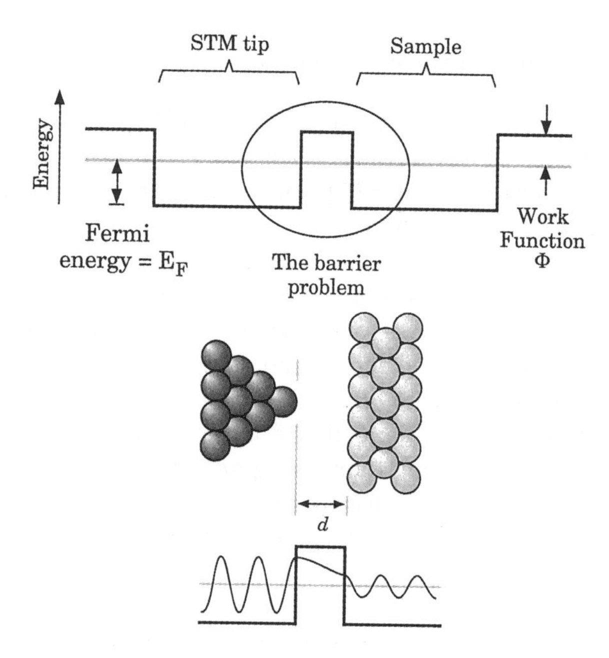

Fig. 3.12 Tunnelling through a potential barrier

and sample as free electron metals we see that we have a potential barrier problem that has become a standard application of undergraduate quantum mechanics (Fig. 3.12).

In this model we assume that the tip and sample have the same work function and that the bias voltage is vanishingly small. Calculating the probability of an electron tunnelling through this barrier requires solutions of the Schrödinger equation for free electrons moving either side of the barrier and electrons moving within the barrier (when the classical kinetic energy is negative). Matching the values and differentials of the wavefunctions at the boundaries and taking the ratios of the squares of the amplitudes of the wavefunctions on either side of the barrier gives an expression for the dependence of the tunnelling current on the tip–sample separation d

$$I \propto \exp(-20d) \tag{3.1}$$

where d is measured in nm. The derivation of this expression can be found in undergraduate text books on quantum mechanics such as Mandl (1992). Although the magnitude of the tunnelling current is typically only ~ nA, the exponential dependence of I on d ensures that the current is very sensitive to small changes in d—a change in the value of d of only ± 0.1 nm will cause I to change by an order of magnitude. This explains the ability of an STM to detect the corrugation of a surface on an atomic scale.

An image of the surface can be obtained by operating the STM is one of two modes—constant height or constant current mode (Fig. 3.13). In constant–height mode the tip is rastered over the surface of the sample with the z piezo crystal keeping the tip at a constant z value, or height. Due to the topography of the surface, the effective value of the tip–sample separation, or barrier thickness, d, will vary and hence the tunnelling current will vary, producing the image contrast. This mode has the advantages of high contrast (due to exponential dependence of I on d) and high acquisition rates (due to the lack of movement of the z piezo). In constant–current mode, the z piezo crystal forms part of a feedback loop that keeps the tunnelling current constant as the tip is rastered over the surface. The electronics of the data acquisition system detects that the tunnelling current has increased (decreased), then the tip is retracted (advanced) until the desired current is restored. The movement of the z piezo is measured and used to produce a topographic image of the surface.

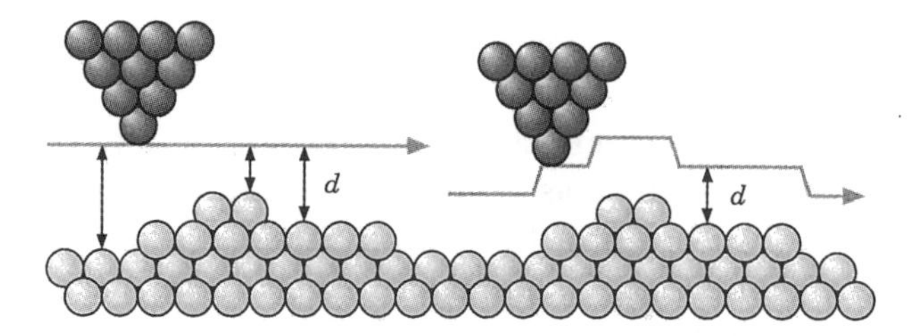

Fig. 3.13 Operational modes of the STM: Constant height (left) and constant current (right).

This mode has the advantage that the tip can move up to avoid contact with features protruding high above the surface, providing that the scan rate is not too fast for the feedback applied to the z piezo. In the field of surface science, this mode has become the standard mode of operation simply because constant–current images provide topographic information in a much more direct manner than do constant–height images. It must be borne in mind that a 'topographic image' is not strictly an image of the

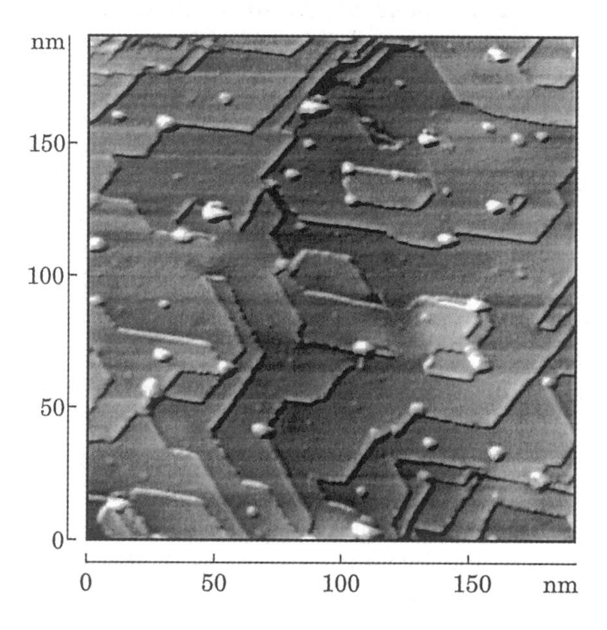

Fig. 3.14 STM image of Sc(0001) taken in topographic, or constant current, mode. The image has been differentiated to give a better impression of the surface topography. Adapted from Dhesi *et al* (1995).

surface of the sample, but an image of a surface that is a contour of constant tunnelling current. For an elemental metallic surface these will be essentially the same, but for surfaces comprising more than one atomic species there may be differences between tunnelling from one atom and another, leading to apparent differences in 'height' for different species of atom. This is particularly true for adsorbates on metallic substrates, where the difference in the densities of states between the adsorbate and the substrate can give rise to substantial variations in apparent height.

An STM image of the (0001) surface of a bulk single crystal of Sc is shown in Fig. 3.14. The image shows flat terraces separated by steps, one atom high, running along the $\langle 11\bar{2}0 \rangle$ directions. These directions correspond to the most densely packed rows of atoms in the (0001) surface plane. The protrusions, that tend to occur at kinks in the steps, are clusters of impurity atoms (carbon and oxygen).

Further Reading

The application of STM and its complementary technique scanning tunnelling spectroscopy (STS) as a local probe of the electronic structure of metal surfaces has been discussed by Wiesendanger (1994). The use of STM in the study of metal surfaces, adsorbates on metal surfaces and metal–on–metal growth has been reviewed by Besenbacher (1996).

3.3 Photoelectron Diffraction

As mentioned earlier, PhD is fundamentally different from LEED or RHEED. The fact that the source of the electrons is internal to the system being studied rather than external to it is not the key feature. The property that differentiates between the behaviour of the electrons in PhD and the behaviour of the electrons in either LEED or RHEED is the coherence of the electron wave. In LEED the incident electron beam has a coherent wavefront over dimensions ~ 10–100 nm, depending on the design of the electron gun. The periodic arrangement of atoms over dimensions comparable with this determine the direction and intensities of the diffracted electron beams. In PhD the incident photon beam is incoherent and each photon creates (at most) one photoelectron, with the result that each photoelectron has a wavefield that is incoherent with that of any others. The origin of diffraction in PhD is in the interference between the photoelectron wavefield and components of the wavefield scattered from

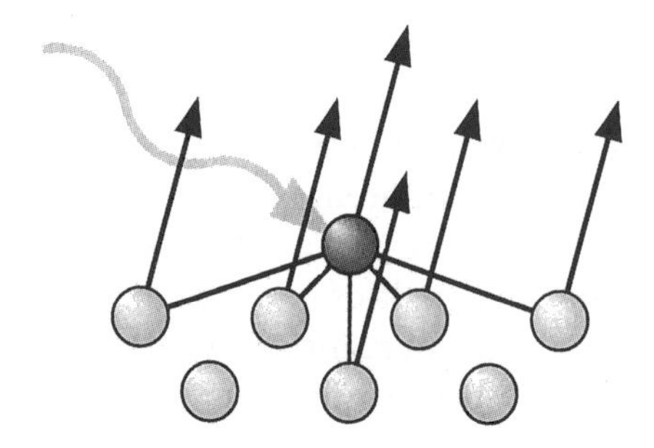

Fig. 3.15 Photoelectron diffraction from an adsorbed atom. The interference between components of the photoelectron wavefield scattering from neighbouring atoms produces photoelectron intensities that vary with emission angle.

atoms neighbouring the photoemitting atom.

As the resultant photoemission intensity is affected by scattering from neighbouring atoms, PhD is sensitive to local atomic structure and so has found application in the determination of the sites and bond lengths of adsorbed atoms. If the photoelectrons have kinetic energies greater than ~ 500 eV then the scattering from surrounding atoms becomes strongly biased towards forward scattering. This is often used to advantage in the determination of the intramolecular axis of adsorbed diatomic molecules, as the maximum intensities of the photoelectron flux from one species of atom indicate the directions of the other species. As the adsorbate atoms will, in general, have core–level energies different from those of the substrate atoms, the adsorbate atoms can be isolated simply by selecting the appropriate photoelectron energies. The same diffraction principles can be applied when there is no adsorbate atom and the sample comprises only one species of atom. In this situation, there is a concomitant loss of surface sensitivity as the photoelectrons may be emitted from any atoms whose distance from the surface is less than a few multiples of the mean–free–path. The resultant diffraction pattern will then be a weighted average of patterns generated by atoms at the surface and in the near–surface region. This is how x-ray photoelectron diffraction (XPD) has been used by Hayoz *et al* (1998) to determine the structure of thin films of Y. Using a Mg K_α x-ray source (photon energy $h\nu = 1253$ eV) the Y $3d_{5/2}$ core–level photoelectrons have a kinetic energy of ~ 1 keV and so their mean–free–paths are ~ 2 nm. This makes the technique sensitive to the local atomic structure around each Y atom in the top ~ 20 layers of the film. As the film was 20 nm thick, this represents a substantial fraction of the full thickness of the film. A diffraction pattern, or diffractogram, from the Y film is shown and discussed in Chapter 5.

Further Reading

The use of XPD in surface structure determination has been discussed by Fadley (1990).

CHAPTER 4

CRYSTAL GROWTH AND SURFACE PREPARATION

4.1 Introduction

The surfaces of the rare–earth metals have been studied since the discoveries of the elements themselves. Many of these studies were not necessarily intended to be surface studies, but were so due to the probing depth of the techniques employed. For instance, attempts to study the electronic band structure of a bulk material using ultraviolet photoelectron spectroscopy will, due to the limited mean–free–path of the photoelectrons, probe only the top few atomic layers of the material.

Early studies of rare–earth metal surfaces used polycrystalline samples due to the difficulties involved with the preparation of single–crystal surfaces of such reactive metals. Surface studies carried out in the 1960s and 1970s gave some information about the properties of rare–earth metal surfaces, but because that information was effectively averaged over all possible orientations of the surface crystallites, possibly weighted by preferential orientation due to surface energy minimisation, it was difficult to draw meaningful conclusions regarding the surface structure or other surface properties. The creation of single–crystal surfaces did not become routine until the 1980s, when the first studies of the geometric and electronic structure of single–crystal rare–earth metal surfaces were published. Studies of the magnetic structure followed shortly thereafter, but were hampered by problems with surface preparation that are only now being resolved.

This chapter looks at the techniques that have been employed to produce rare–earth metal surfaces of the highest quality, in terms of both crystallographic order and chemical purity.

4.1.1 Bulk Samples or Thin Films?

In the study of single–crystal rare–earth metal surfaces there has always been a dilemma. Preparing a clean and well–ordered surface suitable for study involves one of two approaches, neither of which produces the 'ideal' surface discussed in Chapter 2. The first approach is to grow a bulk rare–earth crystal, cut from this a sample having a surface parallel to the desired crystallographic plane using standard metallographic techniques, and then clean the surface of the sample in a uhv chamber. The second approach is to grow epitaxially a thin film of the rare–earth metal on a suitable substrate in the uhv chamber. Each of these approaches will be discussed in detail in the following sections, but the advantages and disadvantages can be summarised rather generally by the schematic surfaces shown in Fig. 4.1.

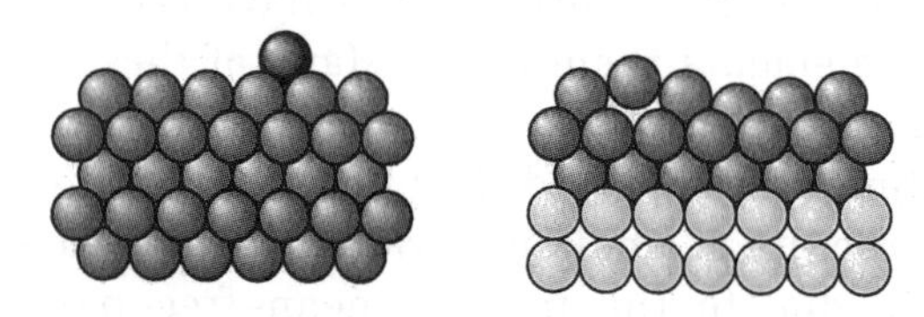

Fig. 4.1 The surfaces of bulk samples have good crystallographic order, but there are always some impurities present (left). By comparison, epitaxially grown thin films have lower impurity concentrations, but generally less well–developed crystallographic order (right). Neither can be considered an ideal surface.

One of the principal restrictions of the epitaxy approach is that the orientation of the crystallographic axes of the thin films grown is determined by the geometry of the unit cell of the substrate, and epitaxial growth of rare–earth metal overlayers occurs only for a limited number of possible orientations. Essentially all of the multilayer rare–earth metal films that have been studied to date have been grown with the close–packed layers of the hcp or fcc crystal structures parallel to the surface. The growth of thin films with surfaces parallel to other planes has been attempted by Du *et al* (1988), but this could be achieved only by using bulk rare–earth single–crystal substrates having the required crystal faces. Thus, for these orientations, the growth of epitaxial thin films did not circumvent the need for the preparation of surfaces of bulk single crystals.

Few experiments on the rare–earth metals, regardless of whether they are bulk–sensitive or surface–sensitive, can hope to provide meaningful information unless the samples used are prepared from the highest–quality material. The reactivity of the rare–earth metals, coupled with the fact that many of the conventional chemical processes that are used to separate elements are not effective for the chemically similar lanthanides, means that purification of the rare–earths is a complex problem. The processes that were developed during the Second World War to extract and purify the actinide element uranium, of obvious strategic importance at the time, were found in later decades to be applicable to the lanthanides.

4.2 Purification and Growth of Bulk Single Crystals

As the developments in the purification of the rare–earth metals and the growth of bulk single–crystals were crucial in enabling the first surface studies, this section outlines the principal techniques that have proved to be successful in the preparation of high–quality rare–earth crystals. The distinction between purification and crystal growth for bulk crystals is somewhat arbitrary because many of the methods employed to chemically purify the material often promote grain growth or recrystallisation at the same time. Section 4.4 will then cover the *in situ* growth of rare–earth thin films by epitaxy.

4.2.1 Purification

Most of the rare–earth material intended for research–grade crystal growth is processed by two collaborating centres specialising in such work — Ames Laboratory, Iowa State University, U.S.A., and the School of Metallurgy and Materials (formerly the Centre for Materials Science), University of Birmingham, U.K. Although the chemical properties of the rare–earth elements are often quoted as being similar, it is often overlooked that their physical properties can vary significantly across the series. For instance, the vapour pressure (at the melting point) varies by 14 orders of magnitude from 10^{-10} Pa (Ce) to 10^{4} Pa (Tm). Such variations in physical properties dictate the most suitable method for purification and crystal growth for each individual rare–earth element. Table 4.1 shows the temperature of the first crystal transformation above room

Table 4.1 Metallurgical properties of the rare earths. Only the first crystallographic transformation above room temperature is shown for each element — for La, Ce, Sm and Yb there are others (see text). The vapour pressure is the value at the melting point.

Element	Crystal Transformation	Transformation Temperature / K	Melting Point / K	Vapour Pressure / Pa
Sc	α hcp → β bcc	1610	1814	8
Y	α hcp → β bcc	1751	1795	0.2
La	α dhcp → β bcc	570	1191	0.00000003
Ce	γ fcc → δ bcc	999	1071	0.00000000009
Pr	α dhcp → β bcc	1068	1204	0.00001
Nd	α dhcp → β bcc	1136	1294	0.002
Pm				
Sm	α Sm → β hcp	1003	1347	600
Eu			1095	90
Gd	α hcp → β bcc	1508	1586	0.01
Tb	α hcp → β bcc	1562	1638	0.09
Dy	α hcp → β bcc	1654	1685	70
Ho			1747	50
Er			1802	40
Tm			1818	18500
Yb	α fcc → β bcc	1068	1092	1900
Lu			1936	0.9

temperature for each of the rare–earth metals (La and Sm both have two transformations — discussed further in the next section), together with the temperature of the melting point and the vapour pressure at the melting point.

The purification of the rare–earth metals may involve a number of stages before the metals can be considered to be of research grade:

(i) vacuum melting — the metals are melted in a vacuum furnace to remove volatile impurities, such as fluorides and hydrogen. For rare–earth metals that have a low vapour pressure at their melting points (such as the light rare–earths, but excluding Sm) this method works well. For rare–earths with a higher vapour pressure, the purification is made difficult because of the loss of metal, even if the furnace is back–filled with an inert gas such as Ar.

(ii) distillation/sublimation — many of the rare earths can be distilled or sublimed to separate the rare–earth metal from the less volatile impurities, such as Ta or W. The rare earths with higher vapour pressures (Sc, Sm, Eu, Dy, Ho, Er, Tm and Yb) which can be distilled or sublimed below 1500 °C can, in addition, be purified with respect to the lighter impurities C, N and O.

(iii) zone refining — a molten zone is moved along the length of a solid rod in one direction several times. Impurities more soluble in the molten metal will move in the direction in which the molten zone is moved, while impurities less soluble in the molten metal will be deposited in the solid metal and will tend to move in the opposite direction. It has been shown that for Y, Ce and Tb zone refining can produce significant purification with respect to transition metal impurities, but the technique is less successful at refining with respect to the interstitial impurities which, in the case of N and O, tend to move in the opposite direction to the transition metals.

(iv) solid state electrotransport (SSE) — a large direct current (up to ~ 300 A) is passed through a solid rod of material for hundreds, or perhaps thousands, of hours. Impurities migrate under the influence of the electric field to one end of the bar, and the speed of the purification process is determined by the magnitude of the electric field that can be applied to the rod. Under uhv conditions (preferred to prevent gaseous impurities from entering the rod) the heat produced by the large current is lost by radiation from the rod's surface. Thus, the field that can be applied is limited by the melting point and vapour pressure of the material and by the diameter of the rod. A small diameter rod has a relatively large surface–to–volume ratio which permits a large electric field to be applied, but which obviously limits the maximum size of crystal that can be produced. The heating of the crystal enhances the electrotransport of impurities along the rod, but also increases the rate of diffusion from the ends of the rod where the impurity concentration is high — the ultimate purity of the rod is determined by the equilibrium reached between electromigration and diffusion. The SSE technique has been used very successfully on rare earths with low vapour pressures, producing metals of higher purity than any other method. It has been found that SSE is very effective for the light elements H, C, N, and O. It has also been found to be effective for Fe impurities, for reasons that are not clear. Collectively, these five elements typically account for over 95% of the impurities in commercially available rare–earth metals.

As a means of indicating total purity, with respect to both interstitial impurities and other lattice defects, the residual resistance ratio (RRR) is defined as the ratio of the resistance of a metal at room temperature to the resistance at 4.2 K, *ie*, RRR = $R_{300} / R_{4.2}$. SSE was first used in the 1960s, under non-uhv conditions, which yielded Y samples with an RRR of ~ 50 (*cf* ~ 10 prior to the purification). In the 1970s, uhv technology enabled higher values to be achieved; Gd was purified by SSE to an RRR of 175 (corresponding to impurity levels of C and O of ~ 100 atomic ppm) (Jordan *et al* 1974). In the 1980s an Y crystal was produced with a maximum RRR value of ~ 1000 (Fort 1987), yielding samples of very high chemical purity and crystalline quality suitable for surface studies. The highest purity rare–earth crystals grown using SSE are shown in Table 4.2.

Table 4.2 Residual resistance ratios (RRRs) of the highest quality (structural and chemical) crystals grown by SSE processing.

Sc	Y	La	Ce	Pr	Nd	Sm	Eu
520	> 1000	260	–	400	120	–	–
Gd	**Tb**	**Dy**	**Ho**	**Er**	**Tm**	**Yb**	**Lu**
800	> 1000	125	90	60	–	–	150

4.2.2 Bulk Crystal Growth

A number of techniques of crystal growth have been developed and adapted for the growth of rare–earth metal crystals. The choice of technique best suited to a particular element is determined by the physical properties of the element and by the crystal structures that the element adopts. The crystal structures adopted by all of the rare–earth metals as a function of temperature are shown in Fig. 4.2.

Note the multiple structural transformations of La, Ce, Sm and Yb, which can cause severe problems during crystal growth. Growing a single crystal of La is very difficult due to the rather slow fcc–dhcp transformation at ~ 300 °C (260 °C on cooling, 310 °C on heating). The highest quality La 'crystal', in terms of chemical purity and lattice defects, was reported by Pan *et al* (1980) to have a high percentage of the α (dhcp) phase, the

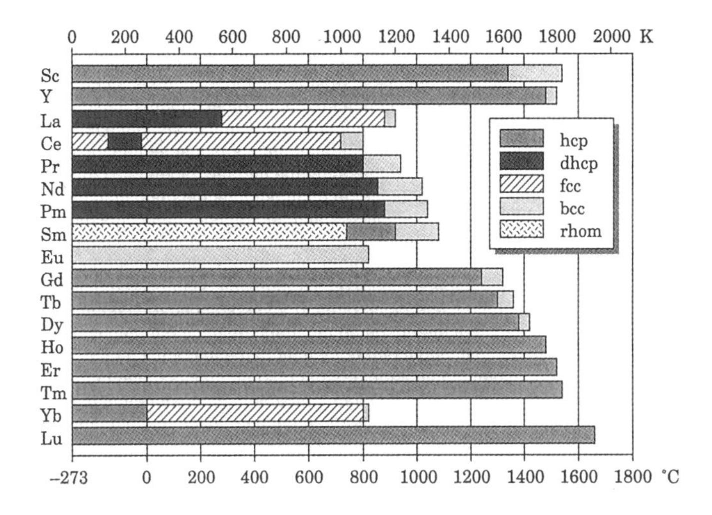

Fig. 4.2 Crystal structure phases of the rare–earth metals.

remainder being the metastable β (fcc) phase. Clearly, such a crystal would be unsuitable for surface structure studies. Below about 200 °C, Ce can exist in three phases: α (fcc), β (dhcp) and γ (fcc). Two of these crystal phases may coexist at a given temperature depending on the thermal treatment of the material (Koskenmaki and Gschneidner 1978). However, under carefully controlled conditions, single crystals of γ-Ce can be produced. Similarly, for Sm a dhcp phase (not shown in Fig. 4.2) may coexist with the rhombic phase at room temperature (Shi and Fort 1987) due to the very small energy differences between these crystal structures. A less obvious cause of problems is the hcp–fcc transition of Yb at ~ 0 °C. Although the transformation temperature may never actually be reached, fcc crystals cooled to room temperature from the melt become highly strained as the transformation temperature is approached. This strain is partially relieved by the formation of a large number of small crystallites, making the growth of large single crystals very difficult. Many workers have reported the coexistence of the α (hcp) and β (fcc) phases due to the slow rate at which the transformation takes place.

There are six principal methods by which rare–earth crystals can be grown. It is worth noting that, even though some methods may produce large crystals of certain elements, these crystals may not be of particularly high structural quality — the growth conditions required to produce the largest crystal size are not always compatible with those required for the highest structural quality. A brief description of each of the crystal growth methods follows:

(i) recrystallisation, or grain growth — an arc–melted sample is annealed at ~ 100 K below its melting point for a period ~ 24 hours to induce grain growth of the polycrystalline material. This has been used to grow large (> 5 cm^3) crystals of Sc and Y, and similarly sized crystals of most of the heavy rare–earth metals.

(ii) vapour deposition, or sublimation — the rare–earth vapour condenses onto a cooled substrate held close to the material (if sublimation is suitable) or crucible (if the material is molten). For rare–earth metals with relatively high vapour pressures (Sm, Eu, Tm and Yb) this has produced crystals a few mm long, though of poor structural quality.

(iii) Bridgman — a Ta crucible containing the melt is slowly lowered through the furnace to induce nucleation at the bottom tip of the crucible and subsequent crystal growth. For rare–earth metals with high melting points there are problems with contamination from the crucible material, forming nucleation sites for new grain growth. The method is most suitable for metals with low melting points.

(iv) Czochralski — a seed crystal, introduced into the melt and slowly drawn upwards, acts as the nucleation point for new crystal growth. The traditional Czochralski method has the same disadvantages as the Bridgman method with regard to contamination from the crucible material. These have been overcome by levitating the melt so that there is no contact between the melt and the crucible. The method has been used to grow crystals of Eu, Y and Gd.

(v) zone melting, or float zoning — a molten or hot zone is moved along the length of a solid rod of metal by localised heating using radio frequency induction or electron beam heating. As Er and Lu do not have a high temperature crystal phase transformation, it has been possible to grow crystals of these metals of ~ 1 cm diameter and 10 cm in length. It has also been used to obtain large Pr and Nd crystals, though due to the dhcp–bcc phase transformation of these elements at ~ 150 K below their melting points, the lengths are somewhat smaller than those of Er and Lu. While the hcp–bcc transformation found in many heavy rare–earth metals close to their melting points does not prevent crystal growth, or seeding for a particular orientation, the final crystal quality is often poor.

A 2 cm^3 crystal of Y (which has such a high–temperature transformation) grown using this technique (Fort 1989) gave poor X-ray Laue diffraction spots, but when subjected to SSE (see below) the structural quality increased significantly.

(vi) solid state electrotransport (SSE) — the SSE technique has already been described as a means of purifying rare–earth metals, but it has been observed that during the purification process the grain size can improve dramatically. In addition to the normal recrystallisation and grain growth processes, the SSE process causes crystalline defects, such as grain or subgrain boundaries, to migrate under the influence of the applied electric field. Crystals grown using SSE generally have one particular orientation parallel to the rod, which is typically ~ 0.5 cm diameter and 15 cm long. If a large crystal of a particular orientation is required then SSE can still be used. A single crystal grown using an alternative method (such as float zoning) can be processed subsequently by SSE to form a large purified crystal. Alternatively, a small single crystal can be oriented and cut in the desired direction with a cross–section that matches the diameter of a polycrystalline rod. The crystal is then butt–welded to the end of the rod and the combined crystal and rod is processed by SSE. This combination of SSE processing with crystals grown using other techniques has resulted in the highest purity rare–earth metal crystals available (Table 4.2).

Further Reading

The purification and crystal–growth techniques that have been applied to the rare–earth metals have been reviewed by Beaudry and Gschneidner (1978a) and more recently by Fort (1987,1989) and Abell (1989).

4.3 Preparation of Bulk Single–Crystal Surfaces

Having produced a high–purity single crystal, the next hurdle for the experimental surface scientist is to obtain a clean and well–ordered surface under uhv conditions — this involves *ex situ* sample preparation and *in situ* cleaning procedures. Producing a sample with a surface having the desired crystallographic orientation can be achieved using the same metallographic techniques that have proved successful for transition metals. The crystal is mounted on a goniometer and the structure and orientation of the crystal is determined using Laue x-ray diffraction (also known as the back–reflection method). Moving the goniometer such that the desired crystallographic plane is vertical, a sample can be cut from the crystal using spark erosion. This is a non-contact method of cutting through a conducting crystal by bringing an earthed wire close to the crystal, which is held at a potential of a few hundred volts. The sparks that jump across the smallest gap between the wire and the crystal vaporise small regions of the crystal. The wire is slowly moved vertically towards the crystal at the rate of a few mm/hour, cutting a slot through the crystal that will expose the desired surface. Keeping the wire and crystal immersed in a bath of a dielectric fluid (such as paraffin) keeps the spark volume small and allows fine control over the accuracy of the cuts made. The surface can then be rechecked with Laue x-ray diffraction, realigned if necessary, and spark–planed by bringing a rotating disk close to the sample surface. This leaves the sample with a surface that is macroscopically parallel to the desired crystallographic plane, but covered in microscopic pits formed by the spark erosion machining. The next step is the removal of these pits and the production of a surface that is microscopically smooth.

Two methods have been developed for the *ex situ* preparation of single–crystal surfaces, one involving electro-polishing and the other relying on mechanical polishing. The choice of *ex situ* preparation has an effect on the *in situ* surface cleaning procedure required, as can be seen by comparing the procedures used for the nine rare–earth metals that have been studied in bulk crystal form.

In the earliest report of a quantitative LEED study of a rare–earth metal, Tougaard and Ignatiev (1982) do not state the origin of their Sc material, nor the details of the *ex situ* surface preparation, but they do describe their *in situ* cleaning procedure for the (0001) surface. The sample was cleaned initially by Kr ion bombardment (at an energy of 1 keV and a sample current density of $20\,\mu A\,cm^{-2}$) whilst the sample was held at a temperature of 600–900 °C. After 15 hours of this treatment, the C, S and Cl contaminants were removed from the surface and the residual O contamination was 5% of a monolayer. Thereafter, a clean and well–ordered surface was obtained by Kr ion bombardment and subsequent annealing at 500 °C for 40 minutes. An earlier Auger electron spectroscopy study of impurity segregation on Sc(100) by Onsgaard *et al* (1979) showed that Ar ion bombardment and annealing at 500 °C gave a surface with considerable contamination: > 25% O, ~ 15% S, < 10% C and < 5% Cl.

During a subsequent anneal, the various contaminants disappeared from the surface at 625 °C (Cl), 680 °C (S) and 840 °C (C and O). However, on cooling to room temperature the S and Cl contaminants returned to their initial levels.

For their photoemission study of Gd(0001), Himpsel and Reihl (1983) electropolished their sample (obtained from Ames Laboratory) in a mixture of 90% methanol and 10% perchloric acid held at dry–ice temperature, using a current of 10 mA cm^{-2}. This process produces a Cl passivating layer that protects the surface from oxidation whilst exposed to ambient atmospheric gases before being inserted into a uhv chamber. Once under vacuum, the Gd surface was cleaned by cycles of Ar ion bombardment and annealing at up to 600 °C. After annealing at this temperature, the sample exhibited a sharp (1 × 1) LEED pattern. It was found that annealing temperatures greater than 700 °C produced a surface with no visible LEED pattern, indicating a surface roughened by sublimation. Comparing the AES peaks of C (273 eV) and O (513 eV) with that of Gd (138 eV) indicated a C–to–Gd ratio of 2–6% and an O–to–Gd ratio of 1–3%, with a 'small' residual Fe contamination.

The report of the photoemission study of Ce(100) by Jensen and Wieliczka (1984) gives little detail of the *ex situ* or *in situ* sample preparation procedures used. The sample was provided already electropolished by Ames Laboratory and many hours of Ar ion bombardment and annealing were required to obtain a clean surface. They note that they had problems with transition–metal impurities (Cu, Ni and Fe) diffusing to the surface. Because of this, the sample was given a final Ar ion bombardment with no subsequent anneal prior to data acquisition. After this treatment LEED spots were still visible, though the sharpness was unspecified, and they assumed that the surface was adequately smooth for studies of the bulk electronic structure.

The Ce(100) sample of Rosina *et al* (1985) was also obtained from Ames Laboratory and electropolished in a methanol/perchloric acid mixture. The *in situ* cleaning procedure comprised Ar ion bombardment (at an energy of 500 eV and sample current of 8 μA for 20 minutes) and annealing at 400 °C. Initially, the surface treated in this way was reconstructed and the photoemission spectra showed hydrogen–induced features. After many cleaning cycles, the bulk hydrogen content was reduced sufficiently that annealing to 280 °C resulted in a (1 × 1) LEED pattern and hydrogen–induced features of negligible intensity in the photoemission spectra. No other contamination was detectable using AES, in contrast to the problems with transition–metal impurities encountered by Jensen and Wieliczka (1984).

The cleaning procedure described by Sokolov *et al* (1989) for their LEED study of Tb and Dy indicated rather different optimum anneal temperatures for each element. The surface of Dy(0001) disordered after annealing at temperatures below 300 °C or above 500 °C. Annealing at 350–400 °C produced 'acceptable' (1×1) LEED patterns, implying that the surface was not particularly well ordered. For Tb(0001), a short anneal of 600 °C produced excellent (1×1) LEED patterns. Attempts were also made to clean Tb $(10\bar{1}0)$, which was found to be much more difficult. After more than 100 hours of Ar ion bombardment and annealing (at unspecified energy, current or temperature) LEED patterns showed a (1×7) or (1×8) pattern that was not very stable.

The surface studies of rare–earth metals carried out at the University of Birmingham and the University of Liverpool (reported in the papers of Barrett *et al* and Blyth *et al* over the period 1987–1992) have used samples prepared using essentially the same procedure for each of the elements studied. The procedure was first reported by Barrett and Jordan (1987) in their angle–resolved photoemission study of Y(0001) and has since been used successfully, with minor modifications, on various crystallographic surfaces of Sc, Y, Pr, Gd, Tb, Ho and Er. The samples were obtained from the University of Birmingham, where commercial material (from Rare Earth Products, U.K., and the Materials Preparation Centre, Iowa State University, U.S.A.) is purified and the rare–earth crystals are grown. After spark–machining the samples to the desired orientation the samples are mechanically polished using a number of grades of diamond paste down to a particle size of 0.25 μm. The surface is left without a passivating layer, and so the sample must be mounted onto a manipulator and inserted into a uhv chamber without delay — for most of the rare–earth metals a time period of a few minutes exposed to atmospheric gases is not a problem. The *in situ* cleaning procedure comprises cycles of Ar ion bombardment and subsequent annealing, usually for a period of 30 minutes each. Where it differs from the procedures already described for Gd and Ce is in the parameters for the Ar ion bombardment — a relatively high energy of 2–3 keV is used if possible, with corresponding sample currents of 10–30 μA cm^{-2}. Ar ion bombardment using these parameters was found to remove the oxide contamination from the surface that forms due to the absence of a passivating layer. The number of cleaning cycles required to produce clean and well–ordered surfaces varies according to the element, crystal face and the actual parameters of the Ar ion bombardment used — Sc(0001) was found to exhibit sharp LEED patterns after only five cycles, whereas many of the lanthanides required 20–40 cycles to achieve reproducibly clean surfaces. The levels of contamination that result from this cleaning procedure are typically < 5% C and <2% O. The advantage of this method over the electropolishing method is that no Cl is introduced into the surface region, and none of the surfaces prepared in this way has shown any sign of contamination with transition–metal impurities. This will be discussed further with reference to studies of Tb.

As yet, Gd(0001) and Tb(0001) are the only rare–earth surfaces that have been prepared using both electropolishing and mechanical polishing methods and have had studies of the resultant surfaces reported. Thus, some comparisons can be made between the preparation methods. Wu SC *et al* (1990,1991) described in detail the preparation method for the surface they used for a photoemission study of Tb(0001). The sample was prepared at Ames Laboratory by electropolishing in a 1–6 vol.% solution of perchloric acid in methanol, with a current density of 500 mA cm^{-2}. *In situ* cleaning comprised Ar ion bombardment (at an energy of 400 eV and a sample current of 2–3 μA for 20 minutes) and annealing at 650 °C (for 10 minutes). Annealing at temperatures above 800 °C caused diffusion of Cl to the surface, resulting in a significantly worse LEED pattern. About 50 hours of cleaning were required to remove the Cl and reduce the concentrations of N, C and O to acceptable levels — the intensity of the AES peaks of C (272 eV) and O (510 eV) relative to that of Tb (146 eV) indicated a C–to–Tb ratio of 3–7% and O–to–Tb ratio of 1–3%. However, they found that their surface was contaminated with a significant concentration of Fe impurity that could not be eliminated — the Fe (703 eV) Auger signal was 10–20% of the Tb signal, indicating an Fe concentration at the surface of 10–15%. A quantitative analysis of the bulk material prior to sample preparation indicated an Fe concentration of 59 ppm, making diffusion from the bulk an unlikely, but possible, explanation for the surface contamination. The Fe contamination observed by Jensen and Wieliczka (1984) on Ce(100), which was also prepared at Ames Laboratory, was at an unspecified concentration. Thus, it is not clear whether or not the source of the Fe contamination was the same in both cases.

The second study of Tb(0001) by Blyth *et al* (1991b) used a crystal, obtained from the University of Birmingham, that was subjected to SSE purification prior to the sample being spark–machined from the boule. The mechanical polishing method, together with the higher energy Ar ion bombardment, produced a clean and well–ordered surface (the optimum anneal temperature used was 650 °C, in agreement with that found by Wu *et al*). Forewarned by the observations of Wu *et al* concerning possible contamination, the presence of Fe on the surface was tested by measuring the photoemission intensity in the region of the Fe 3*p* levels relative to that of the Tb 5*p* levels. No emission could be detected from the Fe 3*p* levels above the noise level of the spectrum, indicating an upper limit of < 1% Fe on the surface. The concentration of Fe impurity in the start material was 4 ppm, and this would be expected to fall during SSE processing. Thus, the Fe content of this sample was considerably less than that of the sample of Wu *et al*, by at least an order of magnitude. Even so, it seems unlikely that the problems encountered by Wu *et al* were a consequence of the Fe contamination of their raw material — it is more likely that the sample was contaminated during preparation.

Hydrogen is often a problem with the rare–earth metals. Invisible to AES, but apparent in photoemission spectra (Barrett 1992a, Hayoz *et al* 1998), H can for some samples be purged from the sample by the routine sample cleaning cycles. For other samples, it has been found to persist even after prolonged cleaning. As H can usually be purged from the rare–earth metals by annealing at modest temperatures (~ 200–300 °C), it is only at room temperature that H induced features are seen in photoemission spectra. Thus, the simplest solution is to acquire data whilst the sample is held above room temperature (giving an additional advantage of a slower contamination rate from the residual gases in the vacuum chamber). Obviously, this can be done only in situations where an elevated temperature is not detrimental to the experiment being conducted.

The anneal temperatures and times that have been used to clean rare–earth surfaces are summarised in Table 4.3. The cleaning procedures used for surfaces other than the close–packed (0001) will be discussed in Chapters 5 and 7 as the surface structure that results can be strongly dependent on the details of the annealing treatment.

Table 4.3 Anneal temperatures and times for surfaces of bulk single crystals. 'n/s' denotes annealing time not specified. 'n/p' denotes values determined by the authors, but not published. Annealing procedures involving different temperature stages are not included here, but are discussed in Chapters 5 and 7.

Element	Surface	Anneal Temp / °C	Time / mins	Reference
Sc	(0001)	500	40	Tougaard and Ignatiev (1982)
		650	30	Dhesi et al (1995)
Y	(0001)	650	30	Barrett and Jordan (1987a)
	$(11\bar{2}0)$	600	30	Barrett et al (1987b,1991b)
La		–	–	–
Ce	(0001)	400	20	Rosina et al (1986)
Pr	(0001)	600	30	Dhesi et al (1992)
Nd		–	–	–
Pm		–	–	–
Sm		–	–	–
Eu		–	–	–
Gd	(0001)	600	n/s	Himpsel and Reihl (1983)
		650	30	Blyth et al (1992)
		1050	2	Quinn et al (1992)
	$(11\bar{2}0)$	600	30	n/p
Tb	(0001)	600	n/s	Sokolov et al (1989)
		650	10	Wu SC et al (1990,1991)
		650	30	Blyth et al (1991d)
		600	n/s	Quinn et al (1991)
	$(10\bar{1}0)$	600	n/s	Sokolov et al (1989)
Dy	(0001)	400	n/s	Sokolov et al (1989)
Ho	(0001)	750	30	Blyth et al (1991c)
	$(11\bar{2}0)$	700	30	Barrett et al (1991a)
	$(10\bar{1}0)$	700	30	Barrett et al (1991a)
Er	(0001)	650	30	n/p
	$(11\bar{2}0)$	700	30	Barrett et al (1991a)
Tm		–	–	–
Yb		–	–	–
Lu		–	–	–

4.4 Growth of Epitaxial Thin Films

Although the preparation of polycrystalline or amorphous thin films of rare–earth metals using *in situ* growth from evaporation sources has been a standard technique for sample preparation for many years, reports of epitaxial growth of thin films did not start to appear until the mid-1980s. During the five–year period 1985–1990 there was a flurry of papers on the structural, electronic and magnetic properties of rare–earth overlayers on various substrates, ranging from submonolayer coverages through to thin films. Of particular note in those early days were the studies of JN Andersen, Chorkendorff and Onsgaard (Denmark), Bauer (Germany), Fäldt and Myers (Sweden), Kolaczkiewicz (Poland), Stenborg (Sweden) and Weller (Germany). The results of these and subsequent studies will be presented in Chapter 5, but here we will cover the basics of epitaxial thin–film growth, including a description of the different classifications of growth modes and how they are identified experimentally.

4.4.1 Growth Modes

The first metal–on–metal epitaxial growth studies can be traced back to the 1930s, but not until the advances in vacuum technology decades later was it possible to analyse and classify the results of experiments in terms of general growth modes (Fig. 4.3). An indication of which of these growth modes is occurring can be achieved using diffraction techniques

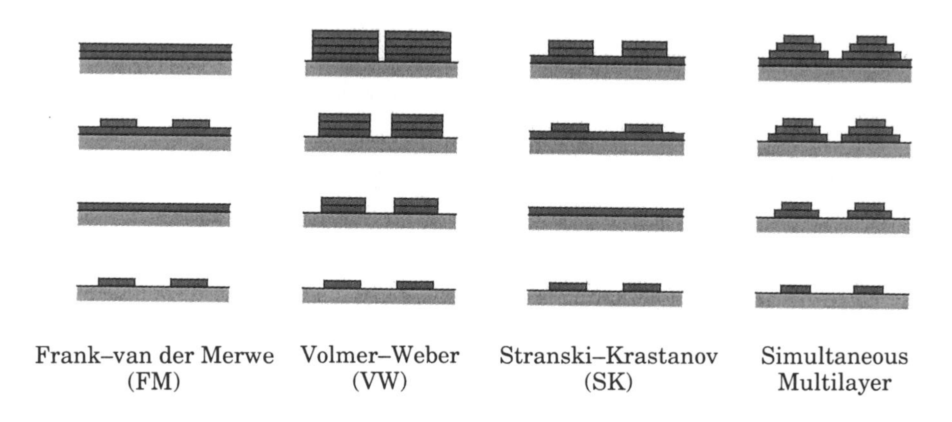

Fig. 4.3 Classification of epitaxial growth modes.

such as RHEED or SXRD, as the intensity of diffracted beams will be dependent on the thin–film morphology. Interference between adjacent layers will be most pronounced if the film adopts the monolayer–by–monolayer (Frank–van der Merwe) growth mode, leading to strong oscillations in the intensity of the diffracted beams, and the other growth modes will produce weaker oscillations. The intensity modulation of the x-rays in SXRD can be compared to predictions based on various possible morphologies. Although RHEED and SXRD are very sensitive to the growth mode, these techniques are not always available to experimenters.

An alternative approach is to use Auger electron spectroscopy (AES) to provide information on the growth mode. Although, by its very nature, spectroscopy is always less direct than diffraction for determining surface structure, AES has the overriding advantage of being a ubiquitous surface science technique. The kinetic energies of Auger electrons are characteristic of the atoms from which they emitted (Fig. 4.4) and so AES has

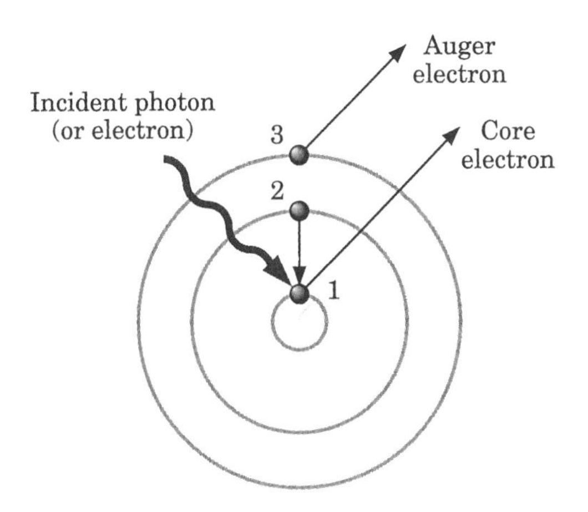

Fig. 4.4 The Auger process: a core electron (1) is ejected by an incident photon or electron; a second electron (2) falls into the resultant core hole; the energy lost by electron (2) can eject a third electron (3).

become a standard technique for identifying and quantifying the chemical composition of a surface. The means by which the core holes are created to produce the Auger electrons is not important for the purposes of determining growth modes — in practice it is usually an electron beam (sometimes from a LEED electron gun), a fixed–energy x-ray source or perhaps a synchrotron radiation source. The growth mode can be inferred by measuring the intensity of electrons at a kinetic energy that corresponds to the Auger electrons from the substrate or the adsorbate atoms. The completion of each layer of the thin film can be identified by monitoring the intensity of the AES signal as a function of time (producing AES(t) plots). Assuming that the flux of adsorbate atoms and their sticking probability are constant with time, then as each layer grows it attenuates the signal from the substrate and adsorbate atoms below it at a rate that depends only on the intensity of the signal reaching that layer. Thus the AES(t) plot shows a series of lines, joined at 'break' points that lie on an exponential curve (Fig. 4.5). The other growth modes shown in Fig. 4.3 will produce AES(t) plots with different characteristic curves, and each of these will now be discussed.

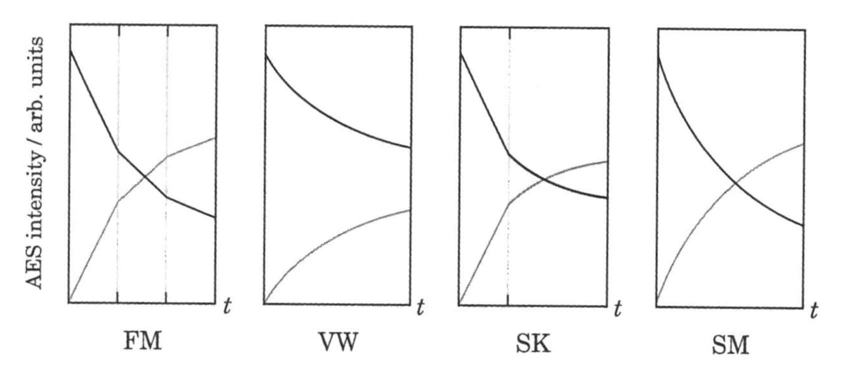

Fig. 4.5 Identification of the growth mode using AES intensities plotted versus time from the substrate (dark) or adsorbate (light) atoms. The tick marks correspond to the breaks in the linear regions of the curves, identifying the times at which layers are completed.

Frank–van der Merwe (FM)

In this mode the growth of the thin film proceeds layer–by–layer, with one layer completing before the next is started. This is an indication of the high mobility of the adsorbate atoms, allowing them to diffuse to the step edges and add to the existing top layer rather than nucleate a new layer. The break points in the AES(t) plot allow the growth rate, and hence the resultant film thickness, to be calibrated to an accuracy of better than 10% of a monolayer.

Volmer–Weber (VW)

No break points appear in the AES(t) plots from either substrate or adsorbate atoms for this growth mode and thus no information can be extracted regarding the coverage of adsorbate atoms. Part of the substrate remains uncovered (until the islands coalesce) and so the attenuation of the substrate signal is not as rapid as for the FM growth mode.

Stranski–Krastinov (SK)

This growth mode is characterised by the formation of one or more layers (as for the FM growth mode) followed by island formation, indicated by a plateau in the AES(t) plot. Adsorbate atoms in islands do not attenuate the substrate signal as effectively as when they form layers, and so the substrate signal attenuation rate is reduced. Also, adsorbate atoms are more likely to intercept the signal from other adsorbate atoms when they form islands and so the adsorbate signal does not rise so rapidly.

Simultaneous Multilayer (SM)

Also known as Poisson growth, this mode results from limited surface diffusion of the adsorbate atoms, promoting the growth of many layers simultaneously. This has the effect of smoothing out the break points seen for the FM growth mode, producing near–exponential AES(t) plots.

The growth mode that is favoured by a particular combination of adsorbate and substrate atom species is determined by the surface free energies of the two materials involved and their interfacial energy. The surface energy of a thin film will depend on the shape and size distribution of islands because edge effects are a function of the film morphology. If such edge effects are neglected then a simple relationship between the surface energies and the predicted growth modes can established. The energy difference between a surface that is completely covered by a film (FM growth) and a surface that is half–covered by islands (VW growth) can be expressed as

$$\Delta E = E_{\mathrm{FM}} - E_{\mathrm{VM}} = (\sigma_a + \sigma_i)A - \frac{1}{2}(\sigma_a + \sigma_i + \sigma_s)A$$
$$= \frac{1}{2}(\sigma_a + \sigma_i - \sigma_s)A \qquad (4.1)$$

where σ_a, σ_s and σ_i are the free energies per unit area of the adsorbate surface, substrate surface and adsorbate–substrate interface, respectively. From (4.1) it can be seen that FM growth is predicted due to its lower energy if $\Delta\sigma = \sigma_a + \sigma_i - \sigma_s < 0$ and similarly VW growth is predicted if $\Delta\sigma > 0$. For the boundary case of $\Delta\sigma \cong 0$, SK growth is favoured. Although crude due to the neglect of island morphology and imperfections such as dislocations or lattice strain, these relationships at least help us understand the origins of the different growth modes. However, as it is not obvious how values for σ_i can be determined for a particular adsorbate–substrate system, their use quantitatively is somewhat limited.

It should be noted that the growth modes described in this section are idealised in the sense that they assume thermodynamic equilibrium conditions. In practice, the growth of an overlayer may occur under conditions of adsorbate flux and substrate temperature such that the equilibrium configuration cannot be reached. Observed growth modes that do not quite fall in to one of the accepted classifications may be described as, for example, pseudo-FM or pseudo-SK.

An example of how AES data can indicate the growth mode is shown

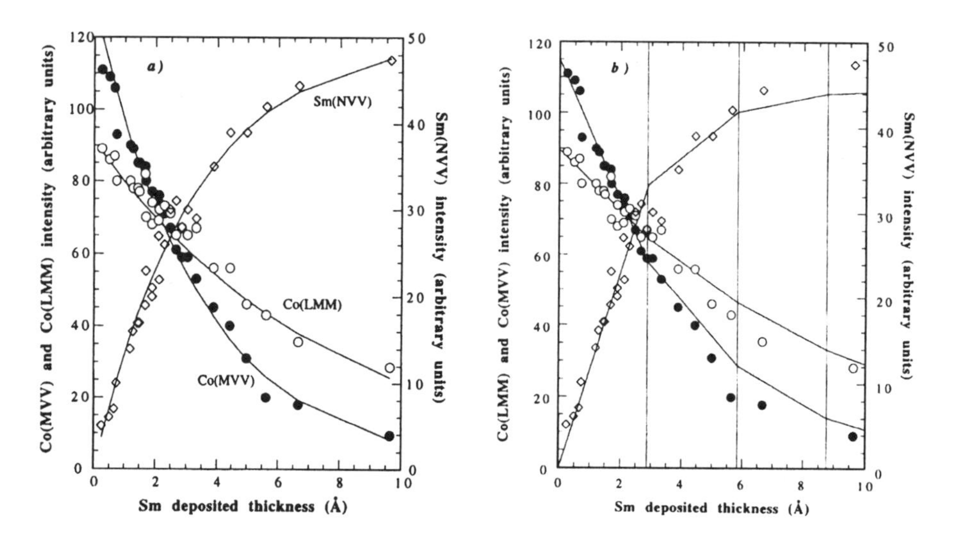

Fig. 4.6 Sm and Co Auger signal intensities are plotted as a function of Sm film thickness (the same data in each graph). The data are fitted to SM (left) and FM (right) growth models. The SM model is clearly a better fit to the data. Adapted from Gourieux *et al* (1996).

in Fig. 4.6 for Sm on Co(0001) (Gourieux *et al* 1996), where the data are fitted to the predictions of two possible growth modes. The lines drawn through the data are derived from models of SM and FM growth in which the intensities of the adsorbate and substrate signals are calculated from the fractional areas of the substrate covered by zero, one, two, *etc*, monolayers of the adsorbate and from the attenuation lengths of the Auger electrons. For both models, the free parameters (such as the rate of monolayer formation) have been adjusted to give the best fit for the initial stages of growth. After the first monolayer, the predictions of the two models diverge and it becomes clear that the FM growth model cannot account for the AES(t) data.

4.4.2 Measuring Film Thickness

If the coverage of adsorbate atoms is limited to just a few monolayers, then observation of the break points in AES(t) plots is a common method to calibrate the average thickness of the overlayer. For thicker films, the attenuation of a signal from the substrate can be used to determine the film thickness, providing that the attenuation length of electrons in the overlayer material is known. The signal can be an Auger peak intensity in AES, or alternatively a photoemission peak intensity in ultraviolet or x-ray

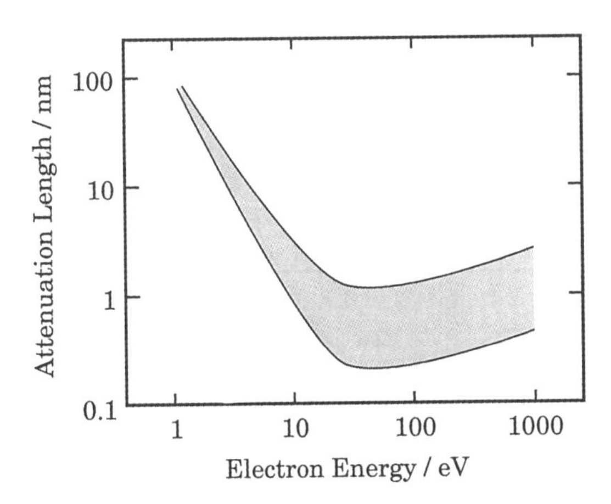

Fig. 4.7 Attenuation lengths as a function of energy. The grey area represents the range of values measured for various materials. Adapted from Seah and Dench (1979).

photoelectron spectroscopy (discussed in more detail in Chapter 5). The attenuation length, often confused with the inelastic mean–free–path, is actually slightly shorter than inelastic mean–free–path because elastic scattering will cause an electron to travel further in total to reach the surface from a given depth. Collecting together the experimental data for a wide range of materials gives the 'universal curve' of attenuation length as a function of electron energy (Fig. 4.7) that provides a good working estimate from which film thicknesses can be derived if the attenuation factor is measured. The values vary by almost an order of magnitude from one element to another and so the curve is very broad, especially at higher electron energies.

Of more use in determining the thickness of rare–earth metal films are the attenuation lengths determined by Gerken *et al* (1982). The values were deduced from measurements of the surface core–level shifts (discussed in more detail in Chapter 5) in photoemission spectra of a number of the rare–earth metals, and cover the electron energy range 20–200 eV (Fig. 4.8).

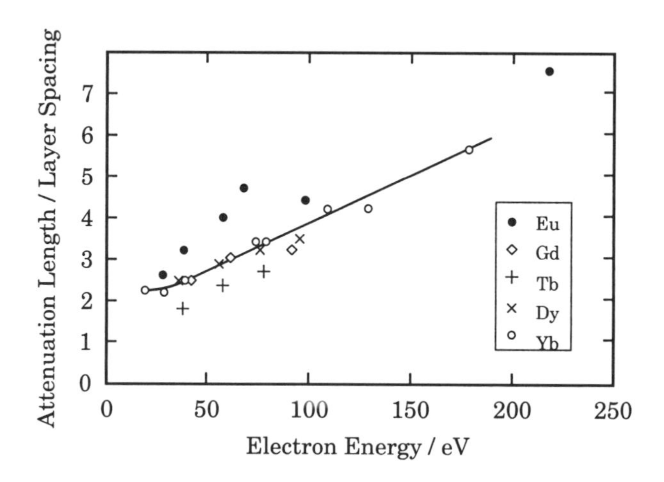

Fig. 4.8 Attenuation lengths as a function of energy for some rare–earth metals. Adapted from Gerken *et al* (1982)

These data indicate that at, say, 100 eV the attenuation length for rare–earth metals is ~ 4 layer spacings. Most of the rare–earth metal (0001) surfaces have interlayer spacings of 0.28 nm (see Table 1.1 and Fig. 1.2), giving an attenuation length ≈ 1 nm. For an AES or photoemission signal that has a signal–to–noise ratio of ~ 1000, a rare–earth overlayer 25 monolayers (7 nm) thick would attenuate the signal to the level of detectability. Over the range ~ 2–20 monolayers this method provides reasonable estimates of film thickness, but as the attenuation is a function of the film morphology, it should be used with caution where accurate thicknesses are required.

In some studies the thickness of the film is not known, but if it is noted that there is no detectable AES or photoemission signal from the substrate, then a minimum film thickness can be calculated as described above.

4.4.3 Growth of Open Surfaces

The structures of multilayer rare–earth thin films will depend on various parameters relating to the substrates — atomic composition, crystal structure and orientation, temperature during the rare–earth deposition, *etc*. Despite all of these variables, if a rare–earth film is grown successfully as a single crystal, then it safe to assume that it will grow with its close–packed planes parallel to the substrate surface. In practice, this means that essentially all of the hcp or fcc films that have been grown have had their (0001) or (111) planes, respectively, parallel to the substrate. This places a severe restriction on the ability of the experimenter to grow films of rare–earth metals with a chosen crystallographic orientation. The only exceptions to the 'rule' of close–packed layer growth have been reported by Du *et al* (1988) and Du and Flynn (1990), each of which employed a rather different approach to the problem.

Du *et al* (1988) noted the strong tendency for hcp rare–earth metals to grow with their *c*–axis perpendicular to the growth plane. Their solution was to use a bulk single crystal of Y, cut to the desired orientation, as the substrate. The surfaces were polished *ex situ* and then prepared *in situ* using the standard method of cycles of Ar ion bombardment and annealing (discussed in the previous section). Surfaces oriented with the hcp *a* and *b* axes perpendicular to the surface, *ie* $(11\bar{2}0)$ and $(10\bar{1}0)$ respectively, were observed by RHEED to undergo a variety of temperature–dependent surface reconstructions, consistent with other studies of these surfaces (Barrett 1992a). The surface reconstructions were not lifted by the homoepitaxial growth of Y from an evaporation source. With the substrate held at ~ 670 K, the RHEED patterns during deposition continued to indicate the presence of substantial surface reconstruction. However, the x-ray diffraction measurements of the resultant films (including Dy/Y superlattices grown on the Y buffer layer) were found to be comparable to those of the substrate crystal, indicating that each reconstructed surface layer must revert to the bulk crystal structure as it is buried beneath freshly deposited material.

Du and Flynn (1990) described a novel crystallographic relationship between hcp rare–earth metals and the (211) surfaces of bcc transition metal substrates. The interface is a low–angle asymmetric tilt boundary that is designed to relieve the long–range strain of the epitaxial film. RHEED patterns indicate that rare–earth metal films more than a few layers thick adopt a strained $(10\bar{1}2)$–type structure and that the motion of dislocations gliding along $\langle 11\bar{2}0\rangle$ directions causes a rotation of the $(10\bar{1}2)$ planes. This rotation will occur if the result is to bring the overlayer and substrate lattices into registry, relieving the overlayer strain and hence lowering the total energy of the overlayer and interface.

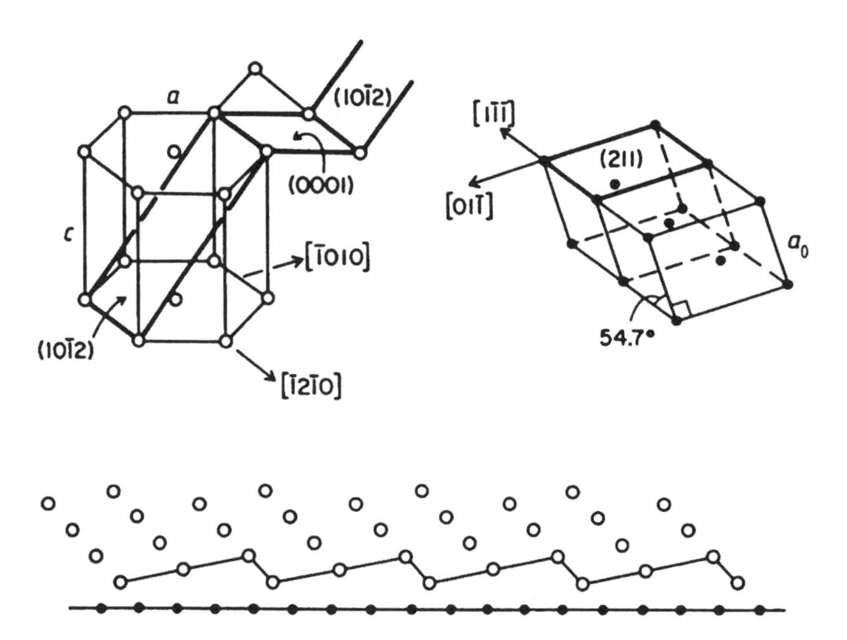

Fig. 4.9 Top left: the $(10\bar{1}2)$ and (0001) planes of the hcp structure. Top right: the (211) surface of the bcc substrate. Bottom: the epitaxial relationship between them as seen from the $[\bar{1}2\bar{1}0]$ direction. From Du and Flynn (1990).

Fig. 4.9 shows a system of $(10\bar{1}2)$ and (0001) planes, formed from $(10\bar{1}2)$ planes that were originally parallel to the bcc (211) surface and now tilted by the introduction of dislocations. Table 4.4 shows the predicted tilt angles calculated for all combinations of hexagonal rare–earth metals grown on refractory–metal substrates.

Table 4.4 Tilt angles in degrees for the growth of the hexagonal rare–earth metals on the (211) surfaces of refractory metals, calculated from the formulae in Du and Flynn (1990).

Rare–earth metal lattice constants / pm				Refractory metal lattice constants / pm			
				330.0	314.7	330.1	316.5
a	c	c/a		Nb	Mo	Ta	W
330.9	526.8	1.592	Sc	8.2°	6.3°	8.2°	6.5°
364.8	573.2	1.571	Y	4.3°	2.1°	4.3°	2.4°
377.4	608.5	1.612	La	2.3°	—	2.3°	0.1°
367.2	591.6	1.611	Pr	3.6°	1.3°	3.6°	1.6°
365.8	589.9	1.612	Nd	3.8°	1.4°	3.8°	1.7°
362.9	582.4	1.605	Sm	4.2°	2.0°	4.2°	2.2°
363.4	578.1	1.591	Gd	4.3°	2.1°	4.3°	2.3°
360.6	569.7	1.580	Tb	4.8°	2.6°	4.8°	2.8°
359.2	565.0	1.573	Dy	5.0°	2.8°	5.0°	3.1°
357.8	561.8	1.570	Ho	5.2°	3.1°	5.2°	3.3°
355.9	558.5	1.569	Er	5.4°	3.3°	5.4°	3.6°
353.8	555.4	1.570	Tm	5.7°	3.6°	5.7°	3.8°
350.5	554.9	1.583	Lu	6.0°	3.9°	6.0°	4.1°

The La/Mo system is different from all the other rare–earth/substrate combinations due to a particular relationship between the lattice constants of the constituent metals. It can be shown that if the condition

$$3a^2 + c^2 < 8a_0^2 \qquad (4.2)$$

is satisfied, where a and c are the lattice constants of the hcp overlayer and a_0 is the lattice constant of the bcc substrate, then two repeat distances of the substrate (measured along $[0\bar{1}1]$, the horizontal direction in the bottom diagram of Fig. 4.9) will exceed one repeat distance of the rare–earth overlayer structure (measured along $[\bar{1}011]$, the long edge of the $(10\bar{1}2)$ plane in the top left diagram of Fig. 4.9). This condition is satisfied for all rare–earth/substrate combinations except La/Mo, due to the fact that La has the largest lattice constant of the rare–earth metals and Mo has the smallest lattice constant of the refractory metals. The lattice–matching condition predicts a tilt angle for La/Mo(211) that is quite different to that predicted for a similar substrate (La/W) or for a similar adsorbate (Pr/Mo) and for this reason the predicted value of the tilt for La/Mo is not included in Table 4.4.

The $(10\bar{1}2)$–type structure has also been observed for Gd/Mo(211) by Waldfried *et al* (1997,1998). Fig. 4.10 shows a LEED pattern from a 2.5 nm thick film of Gd annealed at 650 K. Comparison between the

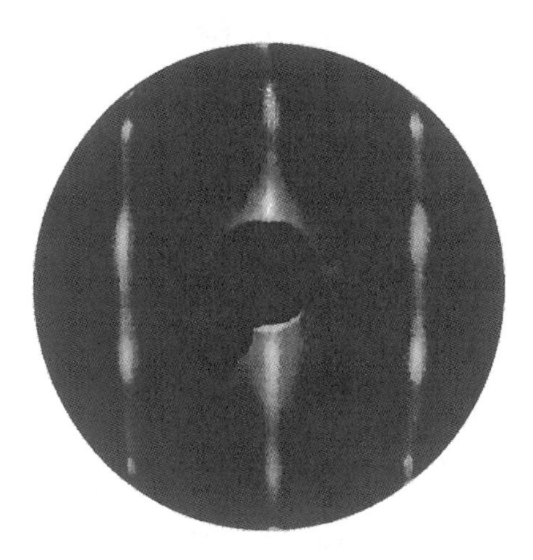

Fig. 4.10 LEED pattern from a 2.5 nm thick film of Gd grown on Mo(211) and annealed at 650 K. The streaks are in the direction perpendicular to the corrugations of the substrate (see Fig. 2.12). From Waldfried *et al* (1997).

positions of the diffraction intensity maxima in this pattern and those of a clean Mo(211) surface indicated that the rectangular unit cell of the Gd overlayer structure has dimensions of 0.374 nm by 0.430 nm. The dimensions of the $(10\bar{1}2)$ unit cell in bulk Gd (the rectangle defining the $(10\bar{1}2)$ plane in Fig. 4.9) are 0.363 nm by 0.855 nm. Thus the film is strained by 3% in the $[\bar{1}2\bar{1}0]$ direction and only 0.6% in the orthogonal

$[\bar{1}011]$ direction. In the latter case, the comparison is made between the dimension of the unit cell of the film and half the value of the corresponding unit cell of the bulk. Contrary to the behaviour observed by Du and Flynn, thicker films (> 10 ML) showed streaked hexagonal LEED patterns, indicating that the films changed to either an hcp (0001) or fcc (111) structure with an in–plane lattice constant ~ 4% greater than that of bulk Gd(0001). There was no evidence of the strain being relieved for films up to 15 nm thick.

The differences between the thicker films of Du and Flynn (1990) and those of Waldfried at al (1997,1998) may have their origins in one of a number of factors. Du and Flynn found that the $(10\bar{1}2)$ tilt angles could deviate from the predicted values, depending on the degree of miscut of the sapphire substrate on which the refractory metal films were grown prior to the growth of the rare–earth metal films (this was followed up by a later study of hcp epitaxial growth by Huang *et al* 1991). Alternatively, the differences may have been the result of different substrate temperatures during growth (500–800 K for Du and Flynn, 150 K for Waldfried *et al*) or from the post–growth annealing of the films carried out by Waldfried at al.

Further Reading

The physical criteria that determine epitaxial metal–on–metal growth have been discussed by Bauer (1982), Bauer and van der Merwe (1986) and Brune (1998), and the epitaxial growth of rare–earth superlattices has been reviewed by Majkrzak *et al* (1991). The use of AES to monitor thin–film growth modes has been analysed and reviewed by Argile and Rhead (1989).

CHAPTER 5

RARE–EARTH SURFACE SCIENCE

5.1 Introduction

In an effort to categorise the hundreds of studies that can be described as rare–earth surface science, Tables A.1–A.4 in the Appendix list the experimental and theoretical studies that have been carried out to date (January 1999). Each of these studies has, to a greater or lesser extent, provided some information about the structure of rare–earth metal surfaces. Tables A.1–A.2 cover studies of the surfaces of bulk single crystals and epitaxial thin films of thickness greater than ~ 1 nm. Tables A.3–A.4 cover ultra-thin films of thickness less than ~ 1 nm. This division at 1 nm thickness is somewhat arbitrary, but reflects the different nature of the experiments. For epitaxially grown films thicker than 1 nm it is often assumed, either implicitly or explicitly, that the substrate has little or no influence on the properties of the rare–earth surface being studied. For ultra-thin films of thickness less than 1 nm (*ie*, less than ~ 4 ML) it is understood that the substrate material and its crystalline structure will have significant influences on the films grown. Some studies include films of various thicknesses that span both regimes, and so these are included in both sets of tables.

Studies of bulk single crystals or 'thick' epitaxial films are tabulated by technique (LEED, RHEED, SXRD or STM) if the specific focus of the study is the surface structure. Studies of the electronic and magnetic structures are labelled under those generic headings, regardless of whether or not information on the surface geometric structure is inferred. Studies of ultra-thin films are tabulated by the substrate material. Refractory metals (usually W or Mo, but sometimes Nb or Ta) are often chosen as substrates if it is the intrinsic properties of the rare–earth films grown that are of interest. For the other substrates, it is the interaction of the rare–earth atoms with those of the substrate, perhaps through interdiffusion, alloy or compound formation, magnetic interaction, *etc*, that is of interest. A significant fraction of the studies of rare–earth metal growth use semiconductor substrates, many with the aim of studying the formation of the rare–earth silicide at the interface with a Si substrate. Most of these studies do not extend our knowledge of rare–earth metal surfaces, and so have not been included in the tables. For readers who want to explore this field further, references to these studies are listed at the end of the book.

The numbers of studies of rare–earth surfaces are shown broken down by rare–earth metal in Fig. 5.1a and by substrate, if appropriate, in Fig. 5.1b. The first graph shows that of the ~ 700 studies reviewed, ~ 25% are of Gd, principally because of its half–filled 4*f* shell simplifying both its electronic structure and its ferromagnetic properties. Apart from Gd, the most–studied rare–earths are Ce, Sm and Yb, which exhibit mixed or intermediate valence behaviour. Many of the studies of these metals have been electron spectroscopy experiments that have studied how the valence of the rare–earth atoms is affected by the presence of a surface or an interface with another atomic species.

The number of materials that have been used as substrates for rare–earth thin–film growth is ~ 50, and so the second graph shows only the most commonly used of those substrates. In addition to the ~ 450 studies shown, there are 21 studies on 8 other transition metals, 23 studies on 10 other semiconductors, 45 studies on other substrates (*eg*, oxides, glass, carbon, stainless steel) and 12 on unspecified substrates. Studies of the growth of one rare–earth metal on another (labelled ‘RE’ in Fig. 5.1b) have focused on the growth on Y (due to its significance in the growth of superlattices) and Gd.

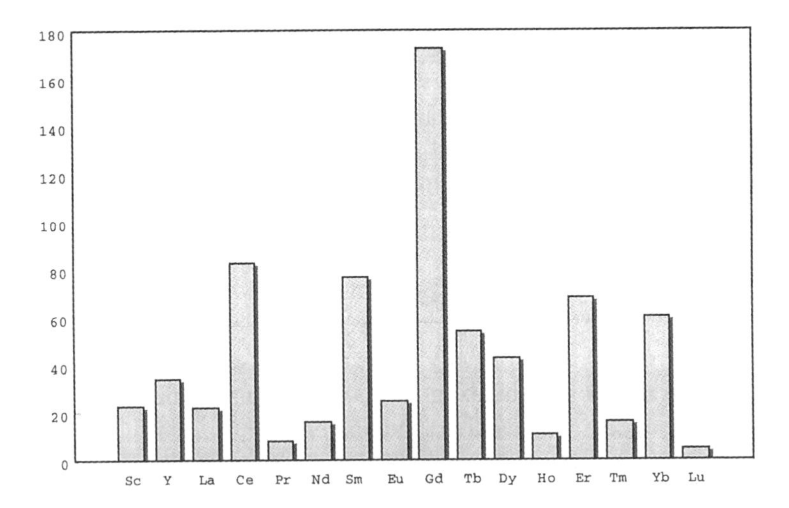

Fig. 5.1a The number of studies of rare–earth metal surfaces. The popularity of Gd, with its half–filled 4*f* shell, is clear.

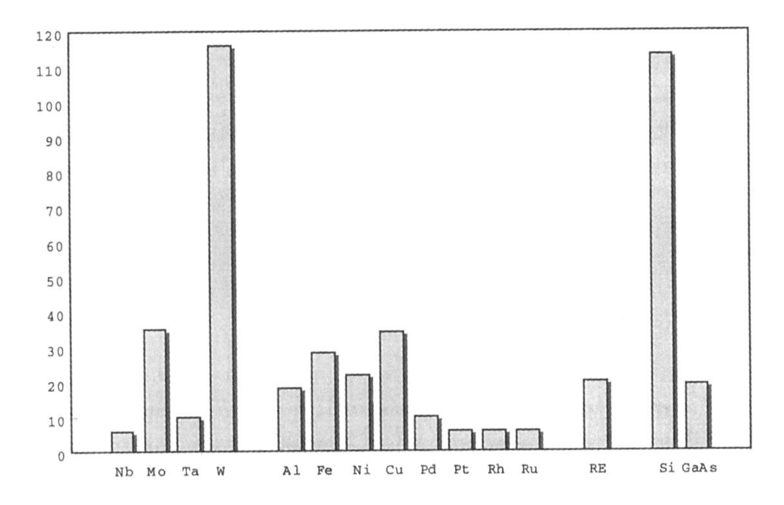

Fig. 5.1b The number of studies of rare–earth metal growth on the most common substrates. The substrates are grouped: bcc refractory metals on the left; fcc transition metals in the middle; rare–earth (RE) metals and semiconductors on the right. Only the refractory metals offer stable and unreactive foundations for rare–earth metal growth — all the other substrates can result in interdiffusion and alloy or compound formation.

The sections that follow in this chapter give an overview of the field of rare–earth surface science, covering experimental studies of the geometric, electronic and magnetic structures and theoretical calculations of surface properties. The first section on geometric structure will, by definition, have direct relevance to our aim of understanding the structure of rare–earth metal surfaces, but the inclusion of studies of the electronic and magnetic structures may at first appear to be of only superficial interest. However, it will become clear in due course just how intimately related the geometric structure is with both the electronic and magnetic structures. A comprehensive review of *all* rare–earth surface science is beyond the scope of this book, but by describing key experiments, some of which would not normally be regarded as providing structural information, we will present the state of knowledge of rare–earth surfaces.

5.2 Geometric Structure

The principles underlying the three principal techniques that have been used to study the geometric structure of rare–earth metal surfaces, LEED, SXRD and STM, have been described in Chapter 3. The application of quantitative LEED requires further discussion — a full description of the theoretical framework required to analyse the results of such studies will been given in Chapter 6 and results will be analysed in Chapter 7. Here, we will describe some of the results of qualitative LEED studies (of which there have been a considerable number) together with RHEED, SXRD and STM studies (which are rather fewer in number).

5.2.1 Qualitative Low–Energy Electron Diffraction

Many studies of rare–earth metal surfaces have involved LEED as an indicator of surface crystallographic quality (in the case of bulk samples and thick films) or of the structure of thin films in the early stages of growth.

5.2.1.1 Surfaces of Bulk Crystals

Qualitative LEED results have been reported for the (0001) surfaces of all of the rare–earth metals that have been studied with surface techniques that are sensitive to surface crystallographic order. In all cases the clean surfaces have displayed (1×1) LEED patterns, as would be expected from stable close–packed surfaces (Barrett 1992a). This is consistent with earlier studies of other metals that adopt the hcp crystal structure — LEED studies of the (0001) surfaces of Be (Strozier and Jones 1971), Re (Zimmer and Robertson 1974), Zn (Unertl and Thapliyal 1975), Ti (Shih *et al* 1976a), Cd (Shih *et al* 1976b), and Co (Berning 1976, Bridge *et al* 1977, Lee *et al* 1978, Berning *et al* 1981) have shown that the clean surfaces do not differ significantly from the ideal structure (providing the temperature is kept well below any bulk crystal transformation temperature). The LEED spots produced by rare–earth metal surfaces contaminated by more than the usual few percent of C and O are more diffuse, but still form a (1×1) pattern. Under the right conditions of annealing, a contaminated surface can have an ordered overlayer structure that produces a (7×7) LEED pattern, with each of the first–order diffraction spots surrounded by a floret of six fainter spots.

Of rather more interest than the close–packed (0001) surface are the more open $(11\bar{2}0)$ and $(10\bar{1}0)$ surfaces. Reports of surface structural studies of such open surfaces on any hcp metals are scarce. A photo-emission study of Ti$(10\bar{1}1)$ at elevated temperatures by Fukuda *et al* (1978) indicated that the surface underwent an hcp–bcc transformation at ~ 80 K below the bulk transformation temperature, but no LEED observations were made. Structural changes have been reported on the $(10\bar{1}2)$ and $(11\bar{2}0)$ surfaces of Co (Prior *et al* 1978, Welz *et al* 1983), but these were at temperatures close to the hcp–bcc bulk phase transformation at ~ 700 K. Although there has been a report of a surface reconstruction of clean Ti$(10\bar{1}0)$ (Kahn 1975) based upon RHEED data, this was later refuted by Mischenko and Watson (1989a) as their LEED study showed no evidence for a reconstruction — it was suggested that the RHEED patterns were misinterpreted, being produced by surface contamination. Similarly, LEED studies of Re$(10\bar{1}0)$ (Davis and Zehner 1980) and Co$(10\bar{1}0)$ (Lindroos *et al* 1990, Over *et al* 1991) indicate that these surfaces are relaxed but unreconstructed. In a conference abstract Hannon and Plummer (1991) reported a LEED study of Be$(11\bar{2}0)$ in which they observed a (1×3) missing–row–type reconstruction, indicating that every third 'zig-zag chain' of atoms (see Fig. 2.4) is either missing or displaced to occupy the bridge between the two remaining chains.

The LEED study of clean Y$(11\bar{2}0)$ by Barrett *et al* (1987b) was the first to show a surface reconstruction of an hcp metal. This reconstruction involves a change of symmetry; the two–fold symmetric ideal $(11\bar{2}0)$

surface reconstructs to a close–packed structure with six–fold symmetry. This reconstruction was found to take place if the sample was cooled from its anneal temperature of ~ 900 K (a part of the surface preparation cycle, described in Chapter 4) at a controlled rate, whereas a faster cooling tended to 'freeze–in' the ideal surface. Fig. 5.2 shows the LEED pattern obtained from Y(11$\bar{2}$0) in its reconstructed state.

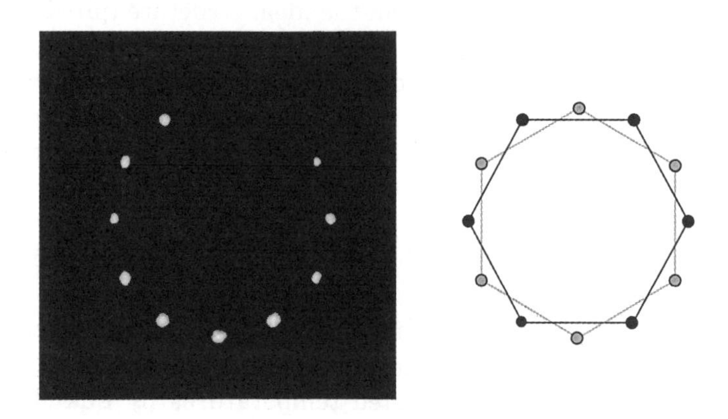

Fig. 5.2 LEED pattern from the reconstructed surface of Y (11$\bar{2}$0) (left) and its decomposition into two sets of spots from hexagonal structures (right). Adapted from Barrett (1992a).

The LEED pattern from the ideal surface (not shown) exhibited the missing spots at $(2n+1,3m)$, where n and m are integers, that are characteristic of a (11$\bar{2}$0) surface (Barrett *et al* 1987b). The 12 principal spots of the reconstructed (11$\bar{2}$0) surface (not all of which are visible in Fig. 5.2 as they are hidden by the sample manipulator) were interpreted as a superposition of two sets of spots rotated by 30° (or 90°) with respect to each other. Each set comprised six spots arranged in a hexagonal pattern characteristic of a (0001) surface. The pattern of LEED spots was slightly larger (by ~ 1%) than those from a Y(0001) sample studied at the same, indicating that the reconstructed surface of Y (11$\bar{2}$0) is slightly more densely packed than a (0001) surface (which is dilated from ideal close–packing by ~ 3%—see Section 1.5.2). Thus, the surface atom density lies midway between the ideal value for close–packing and the value for basal planes of the bulk structure. Tracking the primary electron beam across the sample changed the relative intensities of the two sets of spots, indicating that the surface comprised domains of closely packed atoms, the orientation of which varied over the sample. As the average intensity of the two sets of spots were approximately equal, the two domain orientations were presumed to be equally probable. Fig. 5.3 shows the relationship between the unit cell of the ideal $(11\bar{2}0)$ surface and those of the two domains of the reconstructed surface.

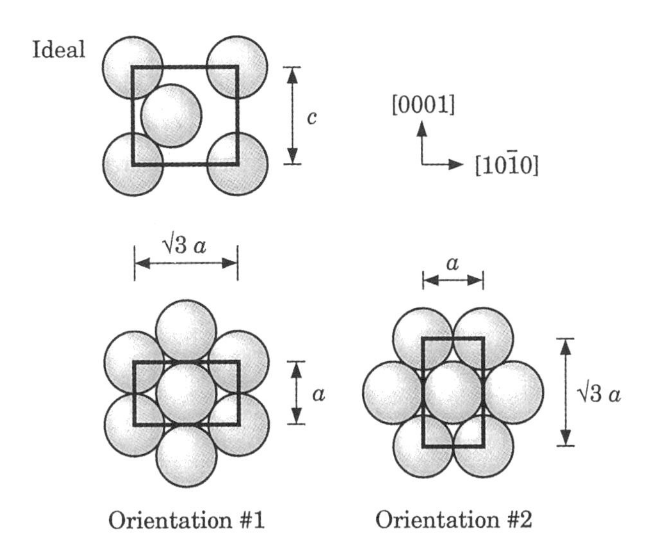

Fig. 5.3 Surface unit cells of the ideal (top left) and reconstructed $(11\bar{2}0)$ surface. Adapted from Barrett (1992a).

The reconstruction can be considered as a contraction of the surface unit cell along either the [0001] or $[10\bar{1}0]$ directions. In the former case (orientation #1), this leads to a unit cell with one dimension equal to that of the ideal unit cell and the other reduced by a factor of $^{c}/_{a}$. The $^{c}/_{a}$ ratio for Y is 1.571 (from Table 1.1), which is noted to be equal to the integer ratio $^{11}/_{7}$ to within 0.1%. For the latter case (orientation #2), the unit cell dimensions differ from those of the ideal cell by ratios of $\sqrt{3}$ and $\sqrt{3}a/c$. Thus, only orientation #1 of the surface unit cell is commensurate with the bulk. The coexistence of both a commensurate and an incommensurate structure means that a Wood notation cannot be assigned to the surface reconstruction.

To determine whether this reconstruction was unique to $Y(11\bar{2}0)$, further LEED studies were carried out. The $(11\bar{2}0)$ surfaces of Ho and Er were found to exhibit only *one* orientation of reconstructed surface, the commensurate orientation #1 (Barrett *et al* 1991a). The spontaneity of the reconstructions for Ho and Er were in marked contrast to the cossetting that was required to allow Y to proceed from the ideal to the reconstructed state, perhaps a consequence of the absence of the two domains. Fig. 5.4 shows the LEED pattern obtained from the single– domain Ho$(11\bar{2}0)$ surface.

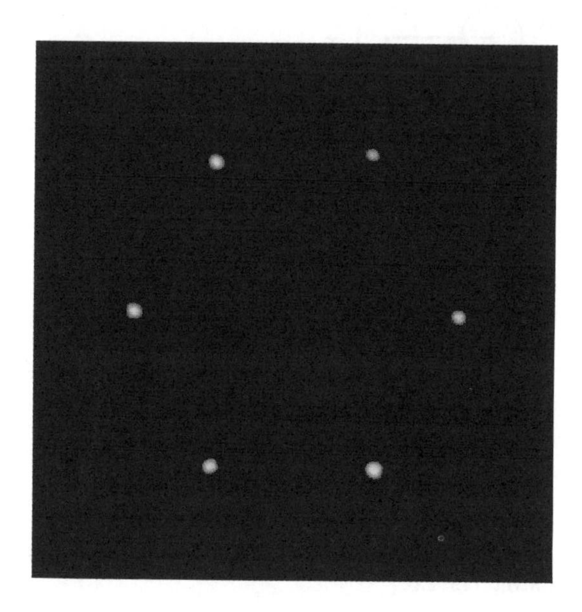

Fig 5.4 LEED pattern from Ho $(11\bar{2}0)$–(7×1). Adapted from Barrett *et al* (1991a).

As this surface unit cell is commensurate with that of the bulk, a Wood notation can be applied to this reconstruction. As can be seen from Fig. 5.3, in the $[10\bar{1}0]$ direction the surface unit cell is lattice matched to that of the $(11\bar{2}0)$, and in the [0001] direction the ratio of lattice parameters is given by the $^c/_a$ ratio of the rare–earth metal. For Y, Ho and Er, the $^c/_a$ ratios are 1.571, 1.570 and 1.569, respectively. The ratio $^{11}/_7$ is equal to 1.5714 and so in each case 11 cells of the surface lattice are an almost perfect match to 7 cells of the bulk. Thus, the reconstruction can be assigned the Wood notation (7×1).

The observation of a single–orientation reconstruction on Ho and Er prompted a reinvestigation of Y $(11\bar{2}0)$, using a different sample from that used in the original study, but cut from the same parent boule. This second study (Barrett *et al* 1991b) showed no sign of the second set of spots that characterised the first. The orientation of the hexagonal LEED pattern showed that, in common with Ho and Er, the surface unit cell collapsed preferentially into orientation #1. The reason for the different behaviour of this second Y sample compared to that of the first is not clear. The surface contamination levels were comparable for both samples, and photoelectron spectra taken with the samples in their reconstructed states were almost identical (covered in a later section). It is possible that the behaviour of the reconstruction is dependent on the precise orientation of the surface relative to the crystallographic axes of the sample — the surfaces were cut to an accuracy of $\sim 1^\circ$, which may not be sufficient to guarantee the reproducibility of the results.

An important observation that sets these reconstructions of the $(11\bar{2}0)$ surfaces of the rare–earth metals apart from all other reconstructions known concerns the depth into the surface to which they extend. The LEED patterns from the $(11\bar{2}0)-(7\times1)$ reconstructed surfaces of Y, Ho and Er showed absolutely no sign of spots from the substrate, indicating a reconstruction depth at least equal to the effective probing depth of a 50 eV LEED beam. This means that the reconstructions of these surfaces are essentially perfect down to a depth of more than five atomic layers. The spots were as sharp as, if not sharper than, those seen from the (0001) surface of any other rare–earth metal. Indeed, the LEED pattern from $Ho(11\bar{2}0)-(7\times1)$ was essentially indistinguishable from that of Ho(0001). The implication is that the reconstructed surface has a degree of crystalline order that is at least as well–developed as that of a (0001) surface, at least parallel to the surface. It seems unlikely that the interface between this structure and the substrate is sharp, and so the conclusion must be that the $(11\bar{2}0)-(7\times1)$ reconstruction comprises a (0001) crystal 'raft', five or more atomic layers thick, sitting on an interface region of indeterminate thickness. The fact that the raft is aligned with the $(11\bar{2}0)$ substrate implies that the interface between them cannot be very thick, but as yet there is no way to determine the structure of this region so far below the surface. Ideas relating to surface recrystallisation from the surface downwards during the annealing of the sample (after Ar ion bombardment has destroyed the crystallinity of the surface region) have been suggested. Although this explains the formation of close–packed surface layers, the problem is then to explain why the surface crystallises in an orientation that is aligned to the substrate far below. If the recrystallisation during annealing occurs in the conventional direction from the substrate to the surface, then the energy required to change the orientation of the crystal structure from $(11\bar{2}0)$ to (0001) must be offset by the lower energy of the (0001) raft. Regardless of whether this reconstruction is an intrinsic property of the $(11\bar{2}0)$ surfaces of the rare–earth metals, or whether is it somehow induced by the presence of impurities or the techniques of surface preparation, it remains a unique reconstruction.

By comparison with $(11\bar{2}0)$, very little work has been done on $(10\bar{1}0)$ surfaces. The only reported studies have been on Tb$(10\bar{1}0)$ (Sokolov *et al* 1989) and Ho $(10\bar{1}0)$ (Barrett *et al* 1991a). In both cases, difficulties were experienced in obtaining clean and well–ordered surfaces and so the results should be treated with caution.

5.2.1.2 Surfaces of Thick Films

For many years it has been predicted that Sm will show unusual surface properties. The bulk metal has a trivalent $4f^5(5d6s)^3$ configuration, whereas photoemission and electron energy loss spectroscopy studies of polycrystalline samples carried out since the late 1970s have shown that the surface has either a divalent $4f^6(5d6s)^2$ configuration or a mixture of trivalent and divalent configurations (see, for instance, Wertheim and Campagna 1977, Wertheim and Crecelius 1978, Lang and Baer 1979, Allen *et al* 1980, Gerken *et al* 1982, Bertel *et al* 1982, denBoer *et al* 1988). The equilibrium radius for divalent Sm atoms is ~ 15% larger than that of trivalent atoms (Rosengren and Johansson 1982) and so the surface layer, if it is entirely divalent, is expected to be expanded relative to the bulk. This was not confirmed experimentally until 1989 when Stenborg *et al* (1989) carried out a LEED and photoemission study of Sm(0001) films. The Sm surfaces were grown epitaxially on a Mo(110) substrate at room temperature. At 80 K the LEED showed a (5×5) pattern indicating a valence–transition–induced reconstruction of the surface layer. The interatomic distances of the atoms in the surface layer were deduced to be larger than those of the bulk by 25% —somewhat more than the value of 15% predicted from the radius of the divalent atoms.

The LEED pattern observed and the corresponding surface structure proposed are shown in Fig. 5.5. Note that the atoms in the surface layer are not close–packed. The surface atom density for this proposed structure

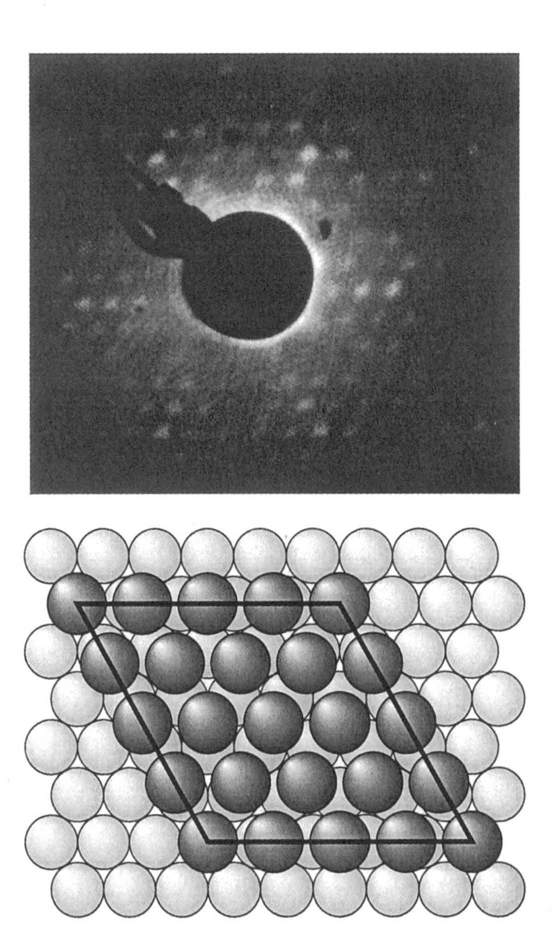

Fig. 5.5 LEED pattern and proposed surface structure for Sm(0001) – (5×5). The darker shading represents the divalent Sm atoms at the surface and the lighter shading the smaller trivalent Sm atoms in the layers beneath. Adapted from Stenborg *et al* (1989).

relative to those of the trivalent bulk layers is $1.25^{-2} = 0.64$. This can be compared to a value of $1.15^{-2} = 0.76$ that a close–packed divalent surface would have. A test for the proposed model was provided by monitoring the relative intensities of the trivalent and divalent features in photoemission spectra, which are well separated in binding energy by many eV. Using the growth characteristics of Sm (Stenborg and Bauer 1987b), where an additional 5% of atoms are incorporated into the first layer during the growth of the second, the ratio of divalent to trivalent intensities showed a sharp maximum at a coverage of 1.62–1.67 ML (where a monolayer is defined as a close–packed bulk Sm layer). This is in agreement with the proposed model, which would predict a maximum in the divalent/trivalent intensity at 1.64 ML.

The (5×5) LEED pattern is observed only at low temperatures; if the Sm is heated from 80 K to room temperature the (5×5) pattern disappears. This reversible order–disorder transition was interpreted as surface melting, indicating a very low surface melting temperature. Such a phenomenon has been proposed (Rosengren and Johansson 1982) as an explanation for the 'anomalously' low melting temperature of bulk Sm metal, based upon data from Spedding (1970). However, more recent data (Beaudry and Gschneidner 1978b) places the melting point of Pm below that of Sm rather than above it, making the quoted anomaly less obvious.

The films used for the study of Sm(0001)–(5×5) were more than six atomic layers thick, such that no signal from the Mo substrate could be detected by AES or photoemission. The precise thickness was not specified, and it is by no means clear what crystal structure the Sm atoms would adopt under such conditions — the rhombic unit cell of bulk Sm metal is nine atomic layers high and so films with a thickness of less than ~ 20 layers (~ 5 nm) cannot be expected to adopt the bulk stacking sequence.

5.2.1.3 Structure of Thin Films

Gd films were studied by Weller *et al* (1985a,1986a,1988) and were grown at a rate of 1–10 pm s^{-1} on a W(110) substrate held at 750 K (as growth at room temperature was found to produce polycrystalline surfaces that gave no LEED patterns). For film thicknesses < 5 nm various LEED patterns were observed (Weller *et al* 1986a). For very low coverages of Gd, patterns were streaked in one direction. As the coverage approached 1 ML, extra spots were visible on a hexagonal lattice — submonolayer coverages gave double diffraction between the hexagonal Gd and the W substrate, which decreased in intensity as the thickness increased up to 5 nm. For thicknesses greater than 10 nm the elastic strain of the Gd film due to the lattice mismatch with the W substrate had been relieved and LEED showed a sharp diffraction pattern of hexagonal symmetry, indicating a well–ordered (0001) surface.

Thin layers of Yb grown on Mo(110) were studied by Stenborg and Bauer (1987a,b) to investigate the effects of valence on the structure. They found that the maximum density of an Yb monolayer corresponded to an atomic radius of 192 pm (*cf* 194 pm for bulk Yb). The proposed model for the surface structure is that of a mesh of closely packed Yb atoms, slightly rotated with respect to the substrate lattice. For both Sm and Yb, coverages beyond a monolayer at room temperature gave LEED diffraction spots that decreased in intensity and sharpness.

Gu *et al* (1991) studied Ce on W(110) at coverages ranging from sub-monolayer to 0.75 nm thick. On increasing the coverage past 0.5 ML a (2×1) LEED pattern changed abruptly to a hexagonal pattern consistent with an fcc (111) structure adopting the NW orientation with respect to the substrate. The in–plane lattice constant started at a value 9% greater than that of γ-Ce, but as more Ce was deposited the lattice constant decreased to a value 9% smaller than that of γ-Ce (and 3% smaller than that of α-Ce). Beyond a monolayer coverage the LEED pattern disappeared, indicating the loss of epitaxial growth.

Continuing the interest shown in rare–earth metals predicted to have valence instability due to the similar energies of the divalent and trivalent electronic configurations, Nicklin *et al* (1992) adsorbed Tm on Mo(110) at room temperature. The submonolayer structure corresponding to their (10×2) LEED pattern was taken to be the same as that proposed for Yb/Mo(110) by Stenborg and Bauer (1987a). At a coverage of 0.5 ML they observed a rapid transition to a pseudo-hexagonal $c(8\times4)$ pattern. This hexagonal structure, contracted by 3% in the substrate $[1\bar{1}0]$ direction, remained during further deposition up to 3 ML.

Room temperature adsorption of Tb, Dy and Er on Mo(110) was studied by Shakirova *et al* (1992). For coverages around 0.5 ML, Tb and Er exhibited weak $c(1\times3)$ LEED patterns and Dy LEED patterns comprised reflections from a mixture of $(1\times m)$–type structures. For all adsorbates, a coverage of 1 ML produced a diffuse hexagonal LEED pattern which disappeared with further adsorption.. Raising the substrate temperature to 800 K gave sharper hexagonal patterns which did not change with increasing film thickness.

Although the (110) surface has been the preferred surface of the refractory metals to serve as a substrate for rare–earth film growth, the effects of other surfaces have been studied. The (211) surface, with its chains of atoms along the $[\bar{1}11]$ direction, has been chosen for studying the effect of the introduction of substantial anisotropy to the substrate topography.

Smereka *et al* (1995) studied the adsorption of La and Dy on Ta(211). Submonolayer coverages of Dy adopted a c(2 × 2) structure and a monolayer coverage resulted in a Dy overlayer with a lattice constant 2% smaller than that of bulk Dy. The situation for La/Ta(211) was rather different, as the lowest coverages of ~ 0.15 ML produced complex LEED patterns and higher coverages led to the loss of the LEED pattern. This behaviour was interpreted as a reconstruction of the Ta substrate. Losovyj (1997) found that no ordered structures were observed for Gd coverages up to 0.6 ML on a Mo(211) substrate, but higher coverages produced c(2 × m) and, over a narrow range of coverages, p(1 × m) structures. The difference in behaviour between Gd and other rare–earth adsorbates on refractory metal substrates (Losovyj *et al* 1982, Gonchar *et al* 1987, Smereka *et al* 1995) was attributed to the interaction between the rare–earth adatoms and the substrates (the electronic structures of the substrates differ in their degree of d-electron localisation).

Waldfried *et al* (1997,1998) observed three phases of Gd film structure when growing on Mo(211). At a coverage of two–thirds of a monolayer a p(3×2) LEED pattern was observed (Fig. 5.6a) corresponding

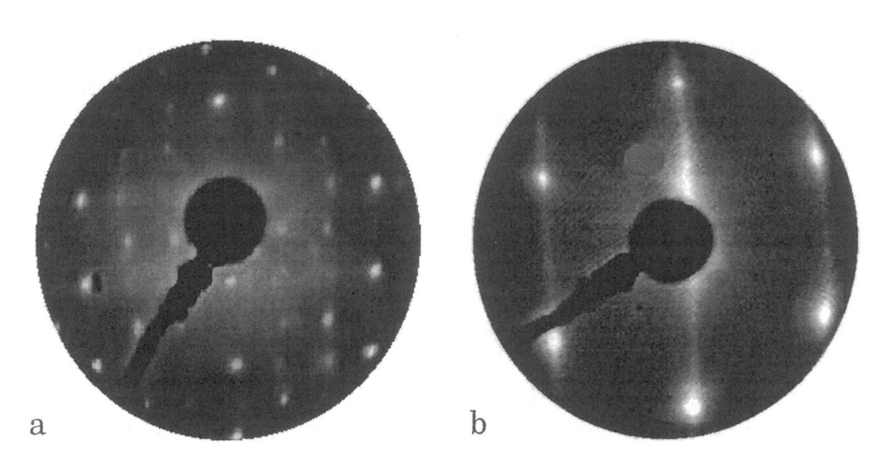

Fig. 5.6 LEED patterns from Gd/Mo(211). (a) p(3×2) pattern from a 0.7 ML Gd coverage. (b) Streaked hexagonal pattern from a 5 nm Gd film. The streaks are in the ⟨110⟩ direction, perpendicular to the substrate corrugations. Adapted from Waldfried *et al* (1997).

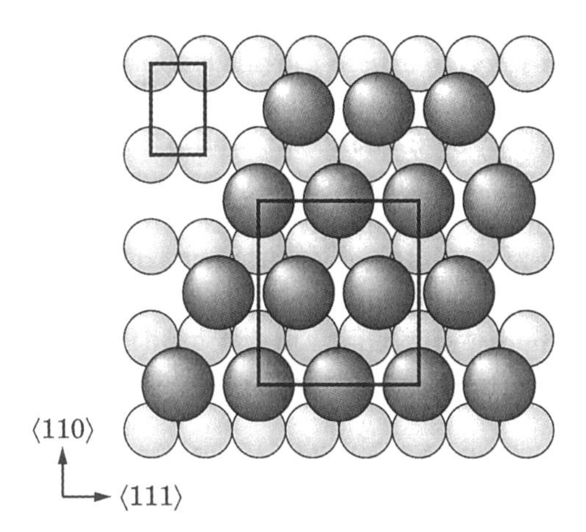

Fig. 5.7 Model for the Gd/Mo(211) submonolayer pseudo-hexagonal structure corresponding to the p(3×2) LEED pattern of Fig. 5.6a. Adapted from Waldfried *et al* (1997).

to the pseudo-hexagonal structure shown in Fig. 5.7. For coverages between 3 and 10 ML, LEED patterns corresponding to the rectangular unit cell of the $(10\bar{1}2)$–type structure were observed. (see Fig. 4.10 and associated text). Gd films thicker than 10 ML formed strained hexagonal structures (Fig. 5.6b) with an in–plane lattice constant ~ 4% greater than that of bulk Gd(0001). Misfit dislocations along the ⟨110⟩ directions of the substrate reduce the long–range order and hence produce streaking of the LEED pattern. The strain was not relieved in films up to 15 nm (~ 50 ML) thick.

Most studies of rare–earth metal growth on refractory metal surfaces have used the (110) or (211) surfaces. The only other surfaces that have been experimented with are the (100) and (111) surfaces.

Thin films of Gd were grown on W(100) to observe the effect of a surface more corrugated than that of W(110) on the overlayer structure (White *et al* 1997). The LEED pattern from a 3 ML film of Gd deposited at room temperature is shown in Fig. 5.8, together with a schematic diagram of the positions of some of the low–index spots. The pattern of spots from the Gd overlayer is readily interpreted as two superposed patterns, each rotated by 90° with respect to the other — a two–domain structure is a

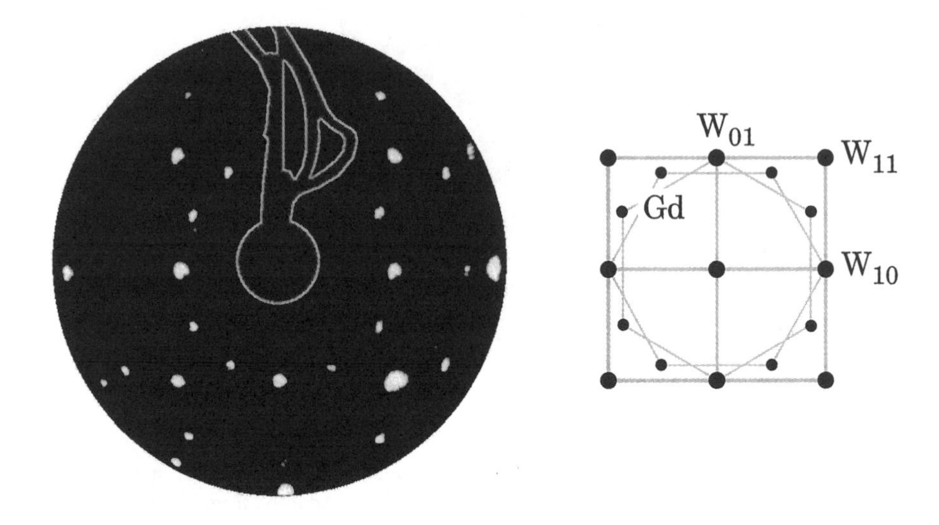

Fig. 5.8 LEED pattern from a 3 ML film of Gd on W(100). The photograph has been overexposed (leading to saturation of the spots from the W substrate) to reveal the fainter spots from the Gd overlayer. The coincidence of some spots in the patterns from Gd and W indicate that the overlayer is commensurate with the substrate. Adapted from White *et al* (1997).

consequence of the four–fold rotational symmetry of the W(100) surface. It is tempting to label the Gd overlayer as hexagonal, but careful analysis of the positions of the spots indicates that some of the Gd spots are coincident with the W spots, implying that the Gd overlayer has a strained pseudo-hexagonal structure that is commensurate with the W substrate. The domains have a c(8×2) periodicity with respect to the substrate, and the proposed model for the structure of the Gd/W interface is shown in Fig. 5.9. The distance between neighbouring Gd atoms measured along the long dimension of the (8×2) unit cell is $^8/_7$ of the W lattice constant. This value is 0.3617 nm, 0.5% less than the bulk Gd lattice constant and 1.3% greater than the value for ideal close–packing (it must be remembered that in bulk Gd the axial ratio $^c/_a$ is 1.591, less than the ideal value of 1.633, and so the atoms in the basal plane are dilated with respect to ideal close–packing by 1.7%). Measured along the short dimension of the unit cell, the distance between Gd atoms is 0.5% greater than in bulk Gd.

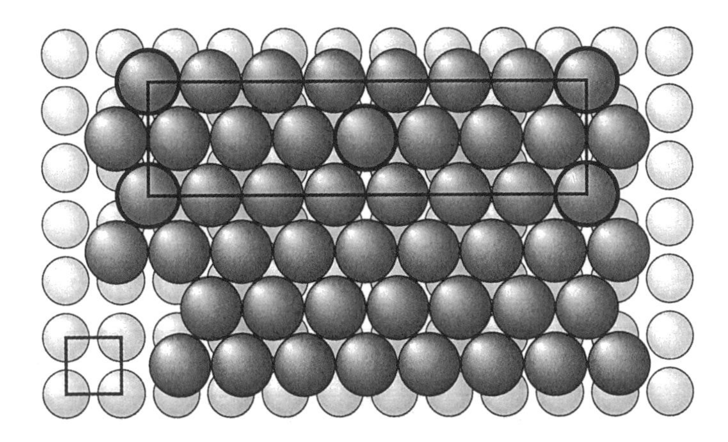

Fig. 5.9 Model of c(8×2) structure of Gd/W(100). The Gd atoms shown in bold outline sit in the four–fold hollow sites of the substrate. Only one of the two domains is shown. Adapted from White *et al* (1997).

In parallel with the studies of rare–earth metals on refractory–metal substrates were studies using transition–metal substrates. Here, the interest was not in the intrinsic properties of the rare–earth metals, but in their interactions with the substrates. In the late 1980s and early 1990s a number of studies were made of rare–earth metal adsorption and overlayer structure. Films of Sm/Al(100) grown by Fäldt and Myers (1985) produced no extra LEED spots at room temperature, but annealing to 420 K produced an ordered c(12×2) structure. Films thicker than 3 ML became unstable, agglomerating on annealing to produce islands of Sm and some regions of uncovered Al. The growth of the Yb/Ni(100) system at 675 K was found by Andersen *et al* (1987) to involve considerable reaction between the Yb and Ni, with the formation of a c(10×2) structure attributed to an intermetallic compound. Dubot *et al* (1990) studied the growth of Sm on the (211) and (110) surfaces of Cr. On the (211) surface the growth was found to be disordered at room temperature, but at a growth temperature of 900 K several successive structures were observed — p(2×2), c(2×2) and p(3×1). On the (110) surface p(2×2) and c(3×3) structures were observed, and then an incommensurate hexagonal structure. Roe *et al* (1994) studied Sm/Ni(111) with an interest in the intermetallic alloys that resulted and Fig. 5.10 shows some of the LEED patterns observed. The Sm atoms in the as–deposited 2 ML film are

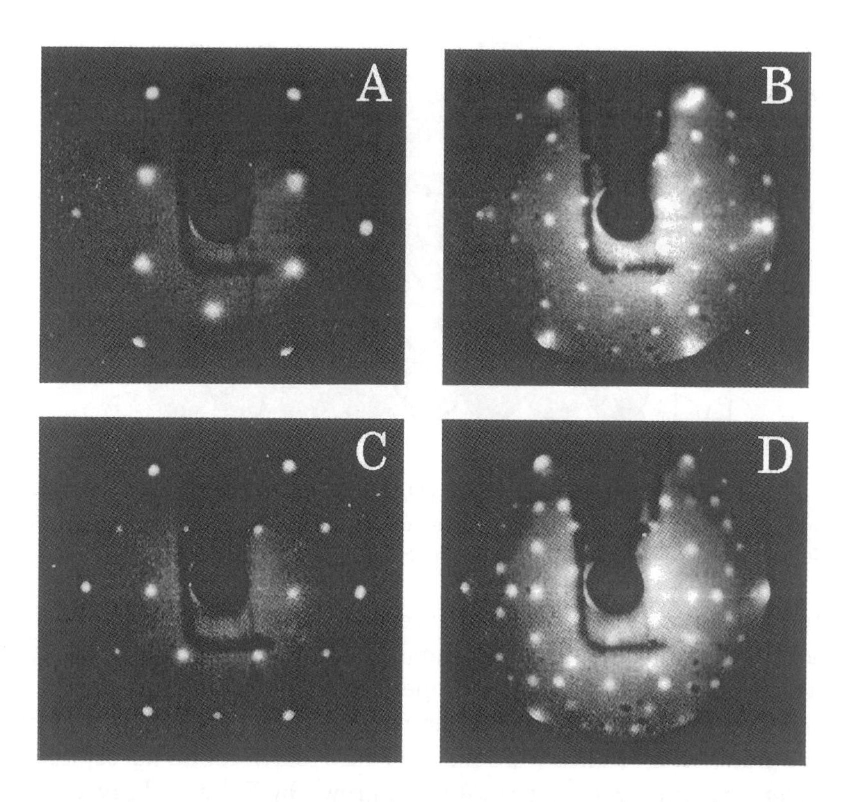

Fig. 5.10 LEED patterns from Sm/Ni(111). (A) a 2 ML Sm film deposited at room temperature, producing a $(\sqrt{3}\times\sqrt{3})R30°$ structure. (B)–(D) various ordered surface alloys produced by annealing 1–4 ML Sm films at 600–800 K. From Roe *et al* (1994).

trivalent, with thicker films showing divalent character. Baddeley *et al* (1997) studied the Ce/Pt(111) system with an interest in the various bimetallic surface phases that resulted from annealing Ce films (of thickness 0.5–4 ML) above 400 K. Deposition of submonolayer films at room temperature resulted in an increase in the background intensity of the (1×1) LEED pattern and the formation of a ring pattern of radius ~ 70% of the separation of LEED spots from the Pt substrate. This was consistent with STM images showing islands of close–packed Ce atoms with random orientation. Above one monolayer coverage the LEED pattern became more diffuse, disappearing at ~ 4 ML. Annealing various coverages to ~ 1000 K produced structures that were ordered surface alloys based on the Pt_5Ce structure.

5.2.2 Reflection High–Energy Electron Diffraction

RHEED has been used in a number of studies of rare–earth metal growth, but not to the same extent as has LEED. Although RHEED is often used in molecular beam epitaxy (MBE) systems, the specialised equipment that is optimised to study epitaxial growth is not as ubiquitous as the LEED optics found in many uhv systems.

Homma *et al* (1987) grew a phase of Ce on V(110) that does not conform to the structures observed for other rare–earth films on bcc metal substrates. The V(110) substrate was grown as a 100 nm thick film on sapphire and the orientation was verified by *ex situ* x-ray diffraction. The 5 nm thick Ce film grew in the FM growth mode with the Ce(111) planes parallel to the V(110) surface. Analysis of the *in situ* RHEED patterns and the *ex situ* x-ray diffraction indicated an in–plane contraction of ~ 8% and an out–of–plane expansion of ~ 2%, respectively, for the Ce film. Thus, this phase of Ce has a trigonal (rather than fcc) crystal structure. It was noted that the lattice distortion does not result in a lattice–matching between the overlayer and the substrate. The orientation of the Ce(111) planes with respect to the V(110) substrate (Fig. 5.11) conformed to neither the NW nor the KS relationships. This orientation is referred to as the R30° relationship by Gotoh and Fukuda (1989) because of the 30° angle between the ⟨110⟩ direction in the fcc overlayer and the ⟨100⟩ direction of

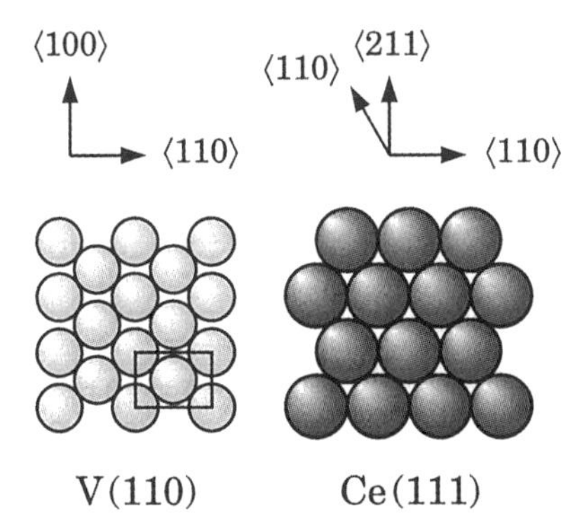

Fig. 5.11 Epitaxial arrangement of Ce(111) on V(110) determined from the analysis of RHEED patterns by Homma *et al* (1987). This orientation is referred to as the R30° relationship by Gotoh and Fukuda (1989). Compare with the NW and KS orientation relationships shown in Fig. 2.13.

the bcc substrate (*cf* 0° and ~5° for the corresponding angles in the NW and KS orientation relationships, respectively). The R30° relationship was predicted by Gotoh and Fukuda (1989) based on an interface energy calculation (IEC) involving the interaction between an fcc (111) monolayer and a rigid bcc (110) surface. They were seemingly unaware of the earlier experimental verification of this orientation relationship by Homma *et al* (1987).

Combining RHEED analysis with IEC (RHEED–IEC), Kamei *et al* (1996) determined the adsorption site of Ce atoms in a monolayer film grown on Mo(110). The resultant structure, not consistent with the R30° orientation relationship predicted by Gotoh and Fukuda (1989), is shown in Fig. 5.12.

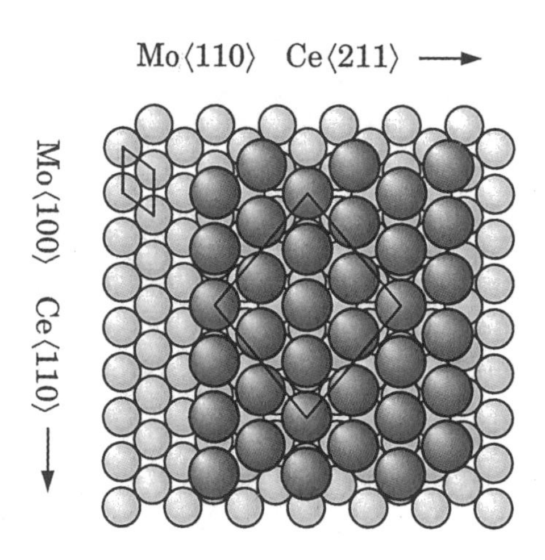

Fig. 5.12 Atomic arrangement and unit cell of the $\begin{pmatrix} 4 & 3 \\ -1 & 3 \end{pmatrix}$ structure of 1 ML of Ce on Mo(110). Adapted from Kamei *et al* (1996).

In a later study, Kamei *et al* (1997) studied the structure of Ce films grown on Mo(100) over the thickness range 0.1–0.6 nm. Five structures were identified with RHEED, and four of these were noted to have the same periodicity (√2 times the Mo lattice constant) along the ⟨110⟩ direction of the Mo substrate.

As for LEED, a number of the RHEED studies of rare–earth films have used transition–metal substrates due to the interest in the interactions between overlayer and substrate.

Ce films on V(110) (Kierran *et al* 1994a) and on Fe(100) (Kierran *et al* 1994b) substrates have been studied with RHEED. In the case of the latter study, the Ce film was found to grow in a disordered state and without interdiffusion with the substrate. Upon annealing at 925 K, an epitaxial CeFe compound was formed.

Gourieux *et al* (1996) studied the growth and structure of Sm films up to 1 nm thick on a Co(0001) substrate. Growth at room temperature of films thicker than ~ 1 ML produced RHEED patterns that were independent of the azimuthal angle between the sample surface and the incident electron beam. On annealing up to 550 K (interdiffusion occurred above 570 K) the RHEED pattern became clearer, but was still independent of azimuthal angle (Fig. 5.13b). The patterns were interpreted in terms of

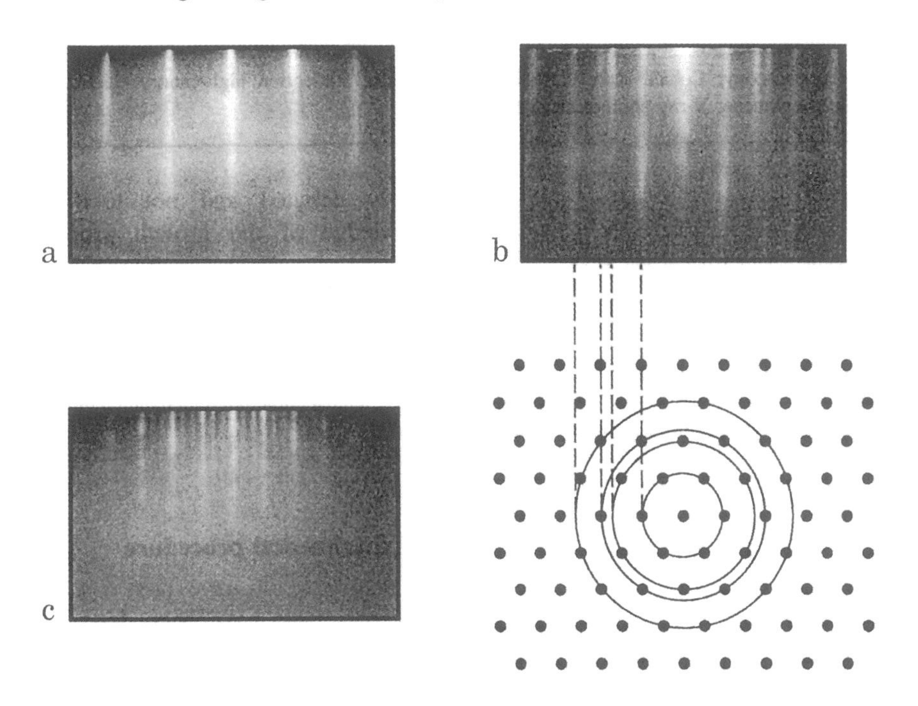

Fig. 5.13 (a) RHEED from Co(0001) at 610 K. (b) RHEED from Sm deposited at room temperature and annealed to 550 K. Below it is the reciprocal space of a polycrystalline two–dimensional Sm layer drawn to scale to show the origin of the different streaks. (c) p(2 × 2) structure observed at 610 K for Sm films thicker than 2 ML. Adapted from Gourieux *et al* (1996).

quasi-two-dimensional polycrystalline layers, with each layer having many facets possessing hexagonal symmetry. The separation of streaks in the RHEED patterns gave an in-plane lattice parameter for the Sm films of 0.377 nm, slightly greater than the value for bulk Sm of 0.363 nm. The faceting of the Sm layers may have been a consequence of the Co(0001) substrate having an increased surface disorder at room temperature compared to the order at the surface preparation temperature of ~600 K. It was noted that the lattice constants of bulk Co and Sm are such that lattice misfits are relatively small for three in-plane orientations of Sm/Co(0001). If all of these orientations occur as facets within a layer, then it would explain the observed RHEED patterns. Growth of Sm/Co at a substrate temperature of 610 K produces a p(3 × 3) structure for Sm coverages up to a monolayer. Growth of the second monolayer caused a structural change that resulted in a p(2 × 2) structure at Sm thicknesses of 2 ML and above (Fig. 5.13c). The persistence of an AES signal from Co, even after the deposition of 50 nm of Sm, indicated that this is an epitaxial intermetallic compound.

5.2.3 Scanning Tunnelling Microscopy

Compared to the use of LEED and other diffraction-based techniques, STM has only recently impacted on the field of rare-earth metal surfaces. Although it has been used to determine the topography of the surface of a bulk rare-earth single crystal (Sc(0001), by Dhesi *et al* 1995, shown in Fig. 3.14) most studies have been of the growth and/or structure of epitaxial films.

In one of the earliest applications of STM to the study of rare-earth films, Tober *et al* (1996) studied the structure of Gd thin films up to 20 ML thick grown on W(110) substrates, with an interest in the relationship between the film structure and the resultant magnetic properties. A monolayer-thick film, annealed at 710 K, was found to adopt a (7 × 14) structure with an in-plane lattice expansion of 1.2% along W ⟨100⟩ and a contraction of 0.6% along W⟨110⟩ compared to the values for bulk Gd. Contrast in constant-height STM images was a result of the corrugation of Gd atoms sitting in various adsorption sites, with maxima in the tunnelling current corresponding to Gd atoms sitting at or close to the positions of W atoms in the substrate below. The same structure was observed for thicker Gd films annealed at 710 K. The Gd formed islands (indicating SK growth) and revealed the (7×14) structure between the islands.

Scanning tunnelling spectroscopy (STS) was put to good use by Trappmann *et al* (1997) to obtain chemical contrast between Gd and Y(0001). STM images from Y films ~100 nm thick grown on Nb(110) at 775 K (Fig. 5.14a) show terraces of width ~50 nm. After depositing ~0.35 ML of Gd on the Y(0001) surface at 475 K, the STM images showed islands of Gd growing on the terraces and stripes formed by Gd atoms migrating across the surface to the Y step edges (Fig. 5.14b). The hexagonal shape of some of the Gd islands is due to the diffusion of adatoms along the high–packing–density edges of the (0001) islands.

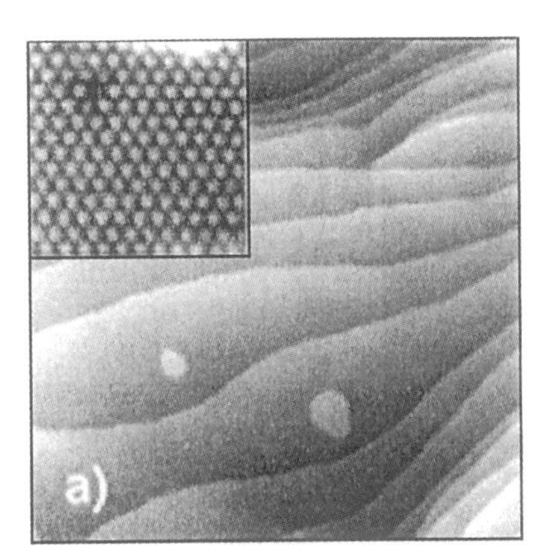

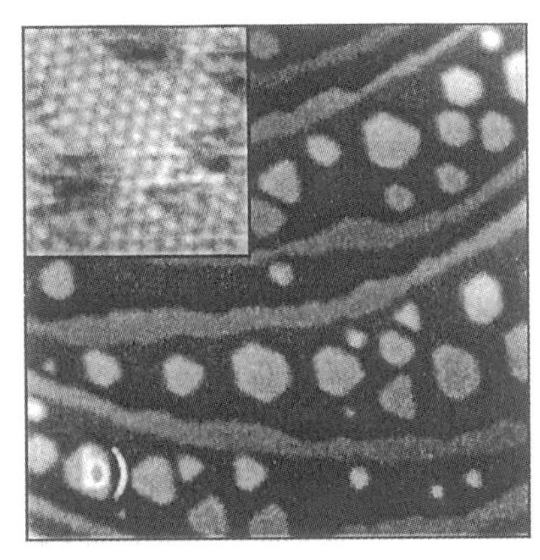

Fig. 5.14 (a) 300×300 nm image of Y(0001) grown on Nb(110). The inset shows the hexagonal basal plane with atomic resolution (~5×5 nm). (b) 100×100 nm image of 0.35 ML of Gd on Y(0001). The inset shows the hexagonal basal plane with atomic resolution (~4×4 nm). Adapted from Trappmann *et al* (1997).

By scanning a series of images using different tunnelling voltages, the contrast between the Gd and Y can be changed and thus the STM can differentiate between the two atomic species (Fig. 5.15). At negative tunnelling voltages (sample relative to tip) the contrast is almost independent of the magnitude of the voltage. At −0.02 V the lighter stripes of Gd are imaged as higher than the Y terraces of the same plane. At +1.2 V the Gd and Y atoms within the same plane appear to be at the same height (and thus the STM image is assumed to be a good representation of the true topography of the surface). At +2 V the Gd is

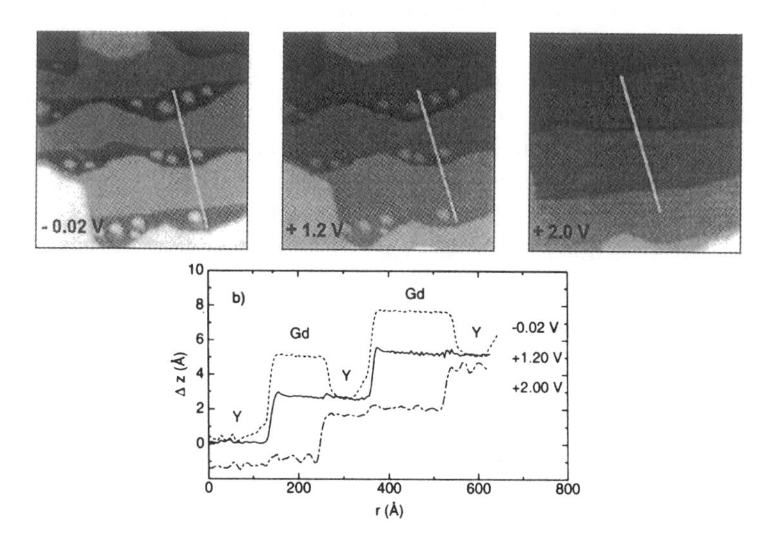

Fig. 5.15 300×300 nm images of 0.8 ML of Gd on Y(0001) obtained at different tunnelling voltages. The white lines indicate the positions of the height profiles displayed in the lower diagram. Adapted from Trappmann *et al* (1997).

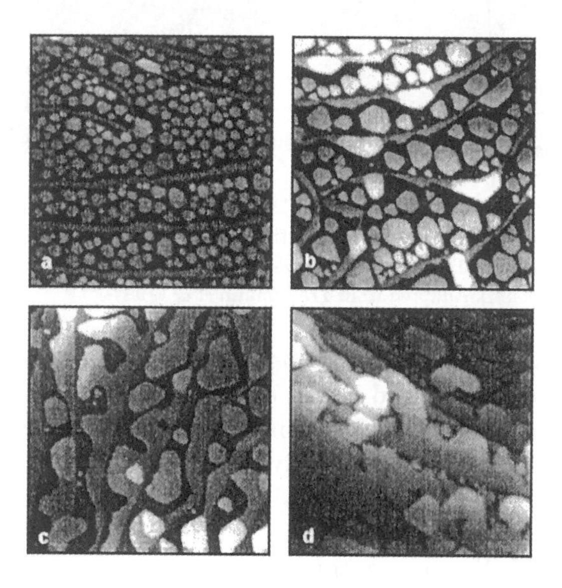

Fig. 5.16 200×200 nm topographic STM images of Gd grown on Y at 473 K. Nominal Gd thickness: (a) 0.11 nm, (b) 0.15 nm, (c) 0.23 nm, (d) 0.29 nm. Adapted from Gajdzik *et al* (1998).

almost invisible, causing a reversal of the image contrast with Gd atoms appearing lower than Y atoms. The sensitivity of the contrast in the STM images to the tunnelling voltages indicates that the images are strongly influenced by electronic effects as well as surface topography. STS was used to investigate the origin of the 'electronic' or 'chemical' contrast and showed distinct *I–V* curves for Y films and Gd islands, some features of which could be identified with electronic surface states above and below the Fermi level. This work on submonolayer films of Gd/Y was continued by Gajdzik *et al* (1998) in a study correlating the structure of the films (Fig. 5.16) with their magnetic properties, as determined by magneto-optic Kerr effect (MOKE) measurements of magnetisation hysteresis loops.

In a series of STM and STS studies of Gd films grown on W(110), Pascal *et al* (1997a–c,1998) reported observing a number of different structures. Submonolayer structures, principally of the $(n\times2)$ type, were imaged (Fig. 5.17) and the local electronic structure probed with STS. For

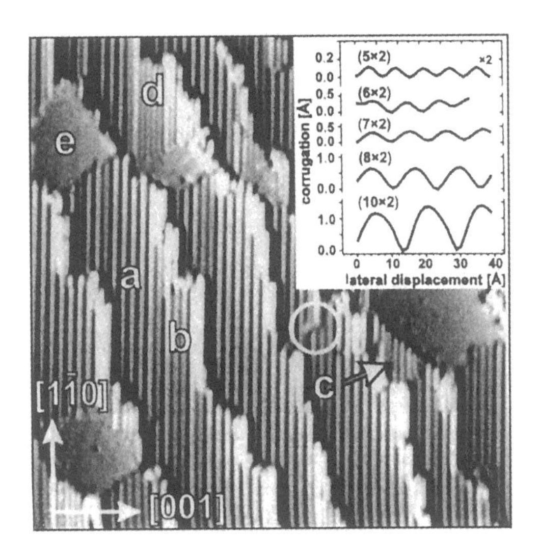

Fig. 5.17 60×60 nm topographic STM image of 0.5 ML of Gd on W(110) annealed at 710 K. Different structures are observed at different locations: (a) (8×2), (b) (7×2), (c) (6×2), (d) (5×2), (e) c(5×3). The inset displays profiles of the different structures along the [001] direction. Adapted from Pascal *et al* (1997a).

the (8×2) and (7×2) structures they were able to distinguish differences in the STS spectra between chains of Gd atoms that lie in either adsorption sites bridging two W atoms or hollow sites between four W atoms.

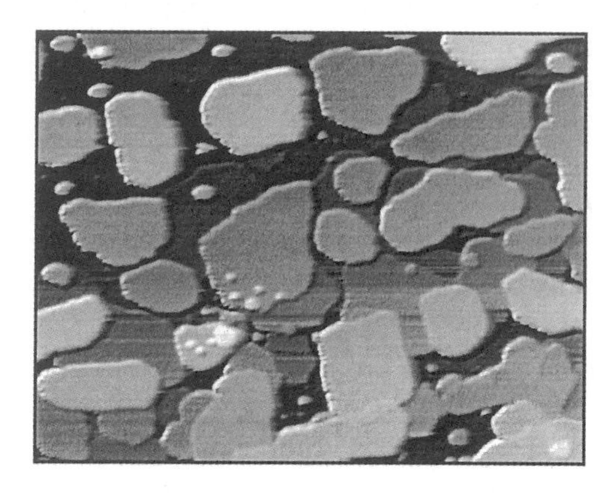

Fig. 5.18 200 × 150 nm STM image of 1.5 ML of Gd on W(100) deposited at 400 K. The multilayer islands rule out FM growth, and second–layer nucleation occurs just before the first layer is complete, indicating a pseudo-SK growth mode. Adapted from White *et al* (1997).

The growth of Gd on W(100) gave rise to a few interesting features (White *et al* 1997). Deposition of submonolayer coverages at room temperature produced small islands ~ 2 nm in size that coalesced upon annealing at ~ 400 K. At a coverage of 0.8 ± 0.1 ML no sign was seen of second–layer nucleation on top of the islands, and continued deposition at 400 K produced larger two–monolayer islands ~ 30 nm in size (Fig. 5.18).

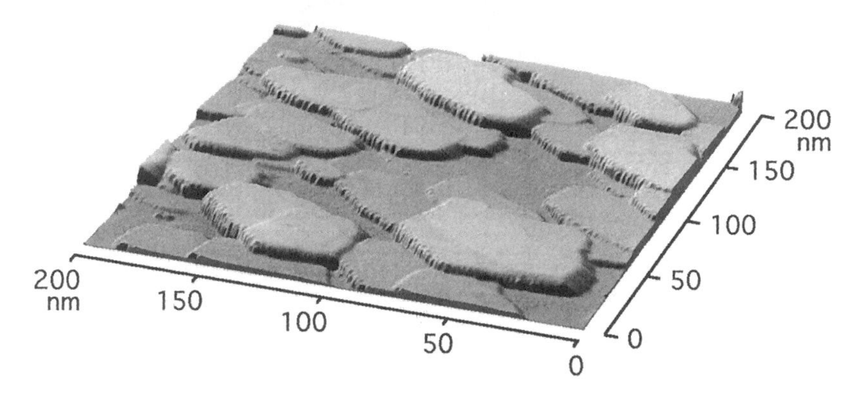

Fig. 5.19 STM image of 1.5 ML of Gd on W(100) annealed at 800 K. At this higher anneal temperature, the islands start to form well–defined edges. Adapted from White *et al* (1997).

Annealing the same coverage at 800 K produced further coalescence of the islands, which started to take on near–hexagonal shapes with relatively straight edges whose internal angles were 120° ± 5° (Fig. 5.19). The surfaces of the islands were observed to be stepped in the opposite sense to the steps of the substrate — where an island traversed a 'down' step in the substrate the surface of the island had an 'up' step and vice versa. This is shown in Fig. 5.20 for a region of the surface found to have a lower density of islands, making the relationship between steps in the island surfaces and those in the substrate more clear. By taking profiles across

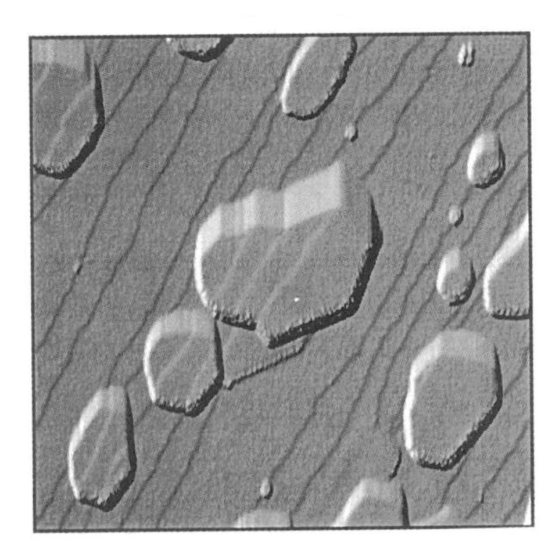

Fig. 5.20 200×200 nm STM image of 1.5 ML of Gd on W(100) annealed at 800 K. By imaging islands that are well–separated from each other, the step structure on the surfaces of the islands can be seen more clearly. Adapted from White *et al* (1997).

islands perpendicular to the step directions, and across areas of the sample devoid of islands, the relationship between the steps could be quantified. For each 0.158 nm 'down' step of the substrate the Gd islands had a ~ 0.13 nm 'up' step. The latter value is close to the difference between the layer spacings for Gd(0001) and W(100) of 0.289 nm and 0.158 nm, respectively, and indicates that this inverted step morphology is a result of the islands reducing the surface energy by minimising the heights of the steps — if the steps in the islands followed those in the substrate, then the step heights would be 0.158 nm, some 20% larger.

Some islands were observed to have a raised structure around the perimeter, imaged as ~ 0.25 nm above the surface of the island (Fig. 5.21).

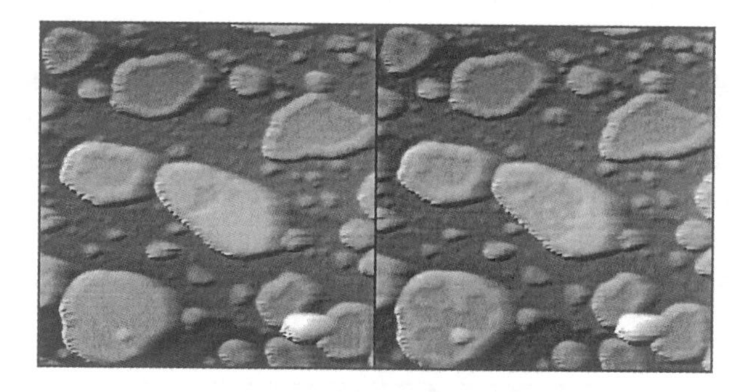

Fig. 5.21 90×90 nm STM images of Gd islands having an apparent raised structure around the perimeter. The islands in the centre and in the lower left were observed to 'collapse' in the five minute period that elapsed between acquiring the two images.

By repeated imaging of an area containing many islands, it became clear that an island started atomically flat and over the course of many minutes (or sometimes hours) a depression would appear, often growing from a nucleation point near the centre of the island, producing the characteristic rampart structure. If an island traversed a step in the W(100) substrate, then a ridge (of a size similar to that measured for the perimeter rampart) would be observed on the surface of the island, running directly over the location of the step beneath (Fig. 5.22).

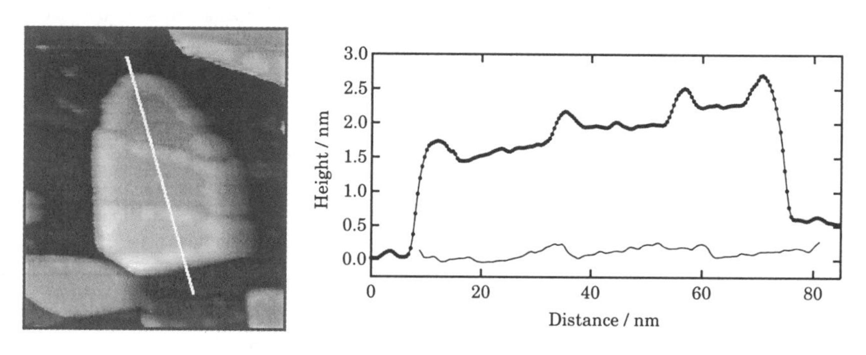

Fig. 5.22 Some islands of Gd on W(100) show ridges running across them corresponding to the positions of the steps in the substrate below. The profile along the white line indicated is shown to the right (line with dots). The structure of the W terraces under the Gd island are indicated by the lower profile (line without dots) taken along a line beside the island (not shown).

These observations were explained by Wiesendanger's group in an STM/STS study of ultra-thin Gd films (4–15 ML, 710 K anneal) grown on W(110) (Pascal *et al* 1997d) that followed their earlier studies of sub-monolayer coverages of the same system (Pascal *et al* 1997a–c). They noted that application of a tunnelling voltage of magnitude greater than 6 V while the tip was located above a Gd island produced a depression 0.2 nm deep. If the area affected was small then the surface relaxed to its initial state on a timescale of several minutes. For larger areas, the effect extended to the whole island and thus the island surface appeared to collapse, leaving the rampart structure referred to earlier. Using STS, the cause of the apparent collapse was traced to the extinction of the surface state of Gd. The surface state is indicated by two peaks in the differential conductivity dI/dV, just above and below the Fermi level. These peaks were present in STS spectra from a flat Gd island, but absent in spectra taken from a 'collapsed' island. The absence of the surface state in the monolayer–thick Gd film between the islands, due presumably to the strain in this film (Tober *et al* 1996), raised the question of whether the island 'collapse' was the result of local lattice strain around the strong electric field at the apex of the tip. Another possible mechanism to explain the extinction of the surface state was material transfer between the tip and sample resulting from the large tunnelling voltage. By exposing a clean Gd surface to ~ 1 L (~ 10^{-4} Pa for 1 s) of hydrogen and noting the change in the STS spectra from a Gd island, and doing the same for ~ 1 L of oxygen, Pascal *et al* concluded that hydrogen was responsible. As

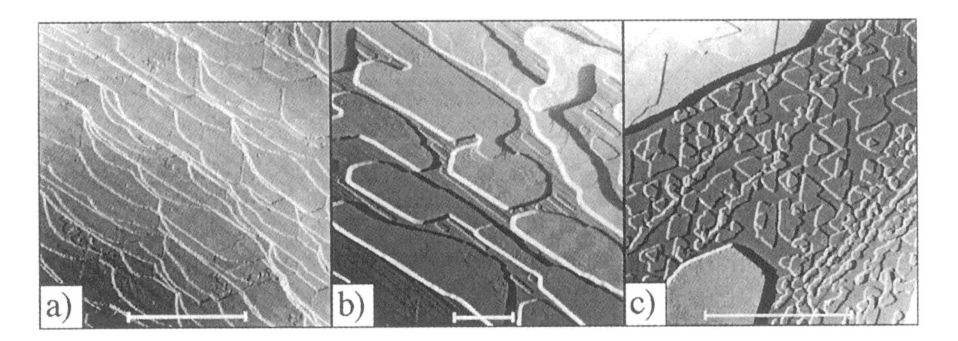

Fig. 5.23 Constant–current STM images of Gd/W(110). (a) Smooth Gd(0001) film with thickness ≥30 ML. (b) Gd islands on a monolayer of Gd/W(110). (c) Two–step prepared sample which exhibits simultaneously local coverages of 1, 2, 3 and >4 ML. The bar scales are 200 nm. Adapted from Bode *et al* (1998a).

hydrogen often represents a substantial fraction of the residual gas in uhv chambers, Gd islands may be expected to 'collapse' spontaneously given enough time, as observed in Fig. 5.21. This work continued with the study of Bode *et al* (1998a), who used STS to study the magnetic exchange splitting of Gd(0001). STM images from the Gd films resulting from various growth conditions are shown in Fig. 5.23. They found that the optimum preparation, based on the work of Aspelmeier *et al* (1994), for smooth Gd films greater than 30 ML thick involved growth on W(110) at room temperature followed by annealing at 700 K (Fig. 5.23a).

The effect of a corrugated substrate on the structure and electronic properties of Gd films was studied by Waldfried *et al* (1996,1997,1998) by using the (211) surface of Mo. In the earlier studies the electronic structure of the strained films were probed with photoemission, and in the latter study STM images (shown in Fig. 5.24) showed the anisotropy of the overlayer structures formed for Gd films 15 ML thick. The domain

Fig. 5.24 STM image of 15 ML film of Gd on Mo(211). The arrow indicates the ⟨110⟩ direction of the substrate. Adapted from Waldfried *et al* (1998).

growth was characterised by long, narrow features (~ 1.5 nm wide) aligned with the ⟨111⟩ direction of the substrate and separated from each other by ~ 2.5 nm. This anisotropic, almost uniaxial, growth was consistent with the streaked LEED patterns referred to in Section 5.2.1.3. Thicker films displayed islands that were similar in shape to those seen on flatter substrates, such as W(110), with typical dimensions of 10–50 nm wide and ~ 50 nm long.

Other STM studies involving rare–earth metals have involved rare–earth/transition metal systems, the magnetic properties of Gd and Tb films sandwiched between other metal layers, and the fabrication of Gd nanowires.

Deposition of low coverages (up to 0.3 nm) of Sm onto Pt(100) (Venvik *et al* 1996) lifted the clean surface reconstruction and produced long and narrow islands, interpreted as a disordered Sm/Pt surface intermetallic compound.

Kalinowski *et al* (1997,1998) measured the magnetic properties of Gd and Tb films of thickness 2–18 nm, sandwiched between W(110) or Y(0001) films of thickness ≥10 nm and capped with 15 nm of Au. The STM was used to determine the surface roughness of the multilayers, which varied over the range 2–15 nm. Clearly, little can be determined from such measurements regarding the structure of the rare–earth film buried 25 nm below the surface.

By annealing a 0.3 ML coverage of Gd on W(110) at 1100 K, Mühlig *et al* (1998) created Gd wires 1 ML high and ~ 5 nm wide. The wires formed at the steps of the substrate on the lower terraces. Annealing the same Gd coverage to 1200 K produced additional islands on the W terraces, which were shown to be Gd atoms embedded in the W surface layer by varying the tunnelling voltage and observing the changes in image contrast. Gd and W are immiscible in the bulk, and so this observation must indicate the formation of a surface alloy.

5.2.4 Other Structural Techniques

The techniques of LEED, RHEED and STM have been used in most of the structural studies of rare–earth metal surfaces, but there are other techniques that have been applied less widely and the results from some of these studies will now be presented.

5.2.4.1 Photoelectron Diffraction

Using ultraviolet (or soft x-ray) photoelectron diffraction Gd overlayers on W(110) have been studied by Tober *et al* (1997,1998). In the earlier study the focus was on the (7×14) interface structure of a Gd monolayer (observed with STM by Tober *et al* (1996) — see Section 5.2.3). A slight excess of Gd metal (1.2 ML) was deposited to ensure the formation of a complete monolayer on annealing to 700–750 K. Due to the core–level shifts resulting from different atomic environments, photoelectron spectra from the W $4f_{7/2}$ levels could be resolved into contributions from W atoms in the bulk of the substrate and those at the interface with the Gd overlayer. Azimuthal PhD scans of the Gd/W interface were found to be very similar to those of the clean W surface. This was attributed to the fact that the Gd atoms sit in a large number of adsorption sites and hence there are a large number of scattering geometries for interface or bulk W $4f$ photoelectrons passing through the Gd overlayer structure. In the later study the focus was on the magnetic properties of a thick (~ 30 nm) Gd film and so spin–polarised PhD (SPPD) was employed. In this technique, a core level is selected that results in a photoelectron spectrum with a well–resolved multiplet of final states that have different degrees of spin polarisation (ideally two final states that have photoelectrons spin–polarised parallel and antiparallel to the moment of the emitting atom). By taking the ratios of the intensities of the photoelectron peaks, and comparing these values with the results of multiple–scattering theory, the magnetic properties of the surface layers (including an elevated surface Curie temperature, to be covered in Section 5.4.1) could be probed. No structural information was derived from the SPPD measurements.

X-ray photoelectron diffraction (XPD) has been used in the study of the effect of hydrogen absorption on the structure of thin films of Y by Hayoz *et al* (1998). With the substrate W(110) crystal held at 600 K, films of Y were grown to a thickness of 20 nm. Fig. 5.25 shows a stereographic projection of the XPS intensities from the $3d_{5/2}$ core level of Y — the centre of the diffractogram corresponds to emission of the photoelectrons along the surface normal and the edge corresponds to emission parallel to the surface. The diffractogram exhibits six–fold symmetry with intensity maxima at angles of 34° (labelled U and W), 50° (X) and 52° (V) with respect to the surface normal. The relatively weak scattering of x-rays means that single–scattering calculations are sufficient to simulate experimental XPD patterns (in contrast to the situation for LEED, where multiple–scattering calculations are required due to the strong interactions between electrons). A single–scattering cluster (SSC) calculation involves building a cluster of atoms around the atom that is the source of the photoelectron and then calculating all possible paths of the outgoing photoelectron that involve zero or one scattering event. Such an SSC

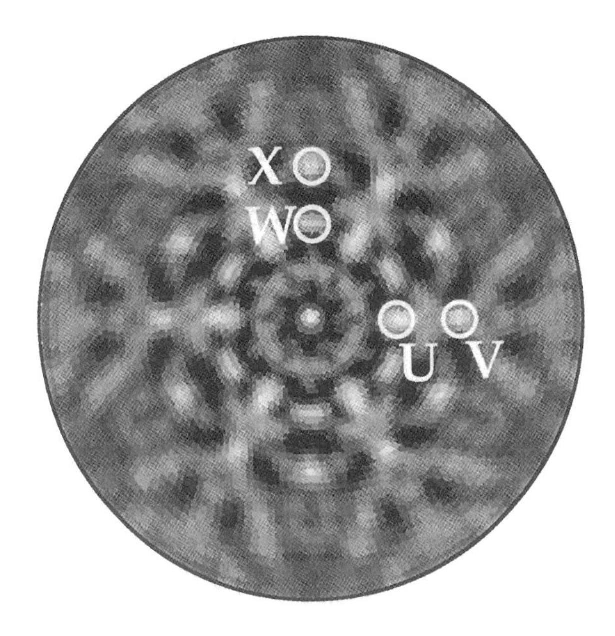

Fig. 5.25 Stereographic projection of experimental Y $3d_{5/2}$ XPS intensities from an as–deposited Y film on W(110). The [001] direction of the substrate is vertical. Brighter areas correspond to higher intensities. Adapted from Hayoz *et al* (1998).

calculation was carried out for Y atoms arranged in the hcp crystal structure with [0001] along the surface normal, and the resultant diffractogram was a close match to the experimental data. The orientation of the diffractogram with respect to the W(110) substrate confirmed that the film grew in the NW orientation (with the $\langle 11\bar{2}0\rangle$ direction of the hcp Y overlayer parallel to the $\langle 100\rangle$ direction of the bcc W substrate). When the Y film was loaded with H, by exposure to 10^{-5} mbar of H_2 at 700 K for two hours, the maxima in the diffractogram labelled W and X disappeared. SSC calculations indicated that the surface comprised two domains of an fcc(111) structure, consistent with the formation of Y dihydride (the H does not contribute to the XPD pattern and so is neglected in the SSC calculations). Annealing the Y film at 1000 K for one hour allowed the film to revert to its original hcp structure. Throughout the cycle of H loading and unloading, the long–range order of the film was confirmed by sharp LEED patterns, implying that the transformation of the clean Y film to the dihydride phase was achieved through the lateral translation of close–packed layers, from an *ABABAB*... to an *ABCABC*.... stacking sequence.

5.2.4.2 Electron Microscopy

In one of the earliest series of studies of the epitaxial growth of rare–earth metals on various surfaces of W substrates, Ciszewski and Melmed (1984a,b,c) used field emission microscopy (FEM) to determine nucleation sites and orientational relationships. In FEM, the substrate is the tip of a sharp needle or a wire etched to a very sharp point, and thus rather different from the flat, single–crystal substrate more commonly used in studies of epitaxial thin–film growth. The tip is placed at the centre of a spherical uhv bulb coated with a fluorescent material. A potential of ~ 10^4 V is applied between the tip and screen such that the electric field lines radiate from the near–hemispherical tip. Due to the small radius of curvature of the tip (< 1 μm), the electric field strength is very high at the tip surface (~ 10^9 V m^{-1}) and can thus ionise surface atoms. The electrons move radially along the field lines and strike the fluorescent screen, producing a real–space image of the distribution of surface atoms magnified by a factor of ~ 10^6. The patterns on this screen can be interpreted in terms of the facets of the crystal structure of the tip.

Crystallites of Gd were found to nucleate on the (110), (100) and (211) facets of the W tip at temperatures of 350–750 K. Nucleation at more than one site resulted in polycrystalline growth, but this could be limited by careful control of the growth conditions; starting with the W substrate at 750 K to allow initial nucleation, raising the temperature to ~ 850 K to avoid further nucleation, and then allowing growth to proceed over a time period ~ 50 min. Both the NW and KS orientations of Gd growth on W(110) were observed, but the KS orientation was found to be more common (most other studies have found NW to be the dominant orientation).

Transmission electron microscopy (TEM) is not a recognised technique for studying surface structure, but as the samples (substrate plus adsorbate) have to be no more than a few nanometres thick to allow the transmission of the electron beam without excessive attenuation, it can be used to determine the crystallographic structure of adsorbate structures. Fuchs *et al* (1993) have used TEM to study Sm clusters grown using low–energy cluster beam deposition (LECBD) onto substrates of Si, glass and amorphous carbon. Deposition at room temperature of Sm clusters (of

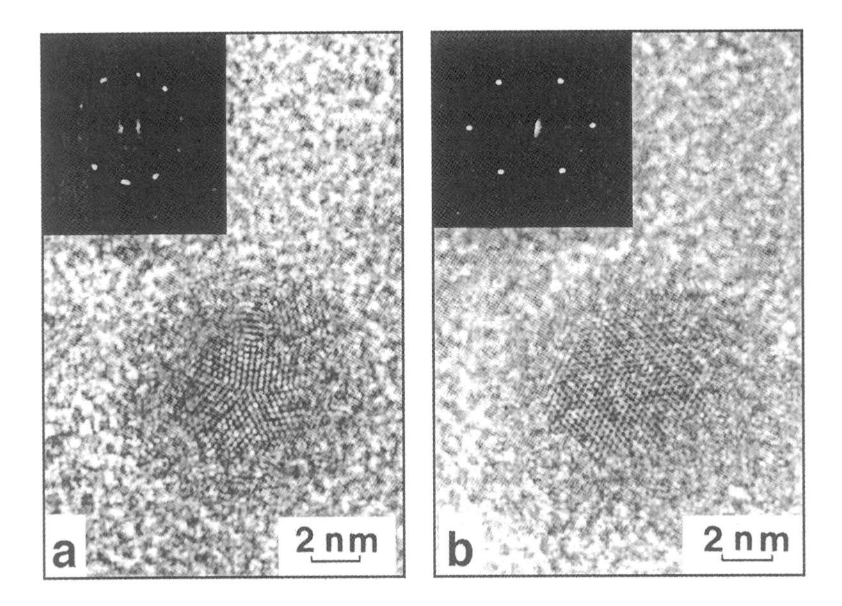

Fig. 5.26 High–resolution TEM micrographs and (inset) corresponding optical diffraction patterns of two shapes of Sm clusters obtained by low–energy cluster beam deposition. The lattice images correspond in both cases to fcc crystallites in the ⟨110⟩ direction. The cluster in (a) is a dodecahedral multiply–twinned particle imaged along a five–fold symmetry axis. Adapted from Fuchs *et al* (1993).

size 3 nm diameter) equivalent to a film of thickness ~ 1 nm produced supported Sm clusters of size 3–7 nm. Doubling the amount of Sm deposited produced twice as many clusters with the same mean diameter, indicating a low–coalescence regime. The crystal structure of bulk Sm is the rhombohedral Sm–type structure (see Section 1.5.2) but analysis of TEM micrographs such as those shown in Fig. 5.26 indicated that the Sm clusters are fcc crystallites. The clusters were deposited in high vacuum (*ie*, ~ 10^{-4} Pa, *cf* ~ 10^{-8} Pa for uhv systems) and yet AES indicated that the clusters were not oxidised significantly during the deposition or by subsequent air exposure. By contrast, Sm films grown by MBE under the same vacuum conditions were clearly oxidised. This led to the conclusion that chemically reactive elements can be grown by LECBD under non-uhv conditions.

5.2.4.3 Medium–Energy Ion Scattering

The surface sensitivity of ion scattering originates from the existence of a region behind a target atom that is shadowed from the incident ions, and thus any other atoms in this 'shadow cone' cannot contribute to the scattering process. If the shadow cones are broad (as is the case for low–energy ion scattering), then surface atoms will shadow all atoms below the surface layer and ensure very high surface sensitivity. For medium–to–high energies (above ~ 100 keV) the shadow cones are more narrow, but surface sensitivity can be maintained by choosing angles of incidence for the ions, relative to the surface crystallographic structure, such that subsurface atoms are shadowed by surface atoms. These 'channelling' directions can be used to derive surface structural information, as a change of surface structure produces a concomitant change in the strength of ion scattering in these directions.

The only medium–energy ion scattering (MEIS) experiment carried out on a rare–earth metal surface was that of Zagwijn *et al* (1997) in a study of the formation of a Sc–Ba–O complex on W(100). MEIS energy spectra and angular distributions were taken after Sc was deposited onto the W substrate and annealed at 1430 K. The peaks in the spectra indicated that the Sc formed a two–dimensional film, and the angular distribution provided a value for the height of the Sc atoms above the W substrate of 191 ± 5 pm (*cf* 158 pm and 291 pm for the layer spacings of W(100) and Sc(0001), respectively).

5.2.5 Growth Modes

The growth modes of rare–earth metal systems have been determined using AES as the principal technique, due to its ubiquity in uhv surface science chambers, with STM becoming more popular in recent years and other techniques playing minor roles. The results of STM studies have been covered in Section 5.2.3, and so in this section the results of some of the pioneering work are described and AES(t) data illustrating the interpretation of the growth modes are shown.

In the late 1980s and early 1990s there were many studies of growth on the close–packed (110) surfaces of refractory metals. Eu, Gd and Tb were grown on W(110) at room temperature by Kolaczkiewicz and Bauer (1986). They found that, up to at least 3 ML, Gd grows in a FM mode that is not thermodynamically stable — upon annealing to 1200 K, films of more than 1 ML thickness agglomerate into large three–dimensional crystals, indicating that the stable growth mode is the SK mode. Weller and Alvarado (1986a) held their W(110) substrate at 725 K for the growth of Gd films. Their AES(*t*) data showed a steep linear increase in intensity of the Gd Auger peak with evaporation time, which then flattened out quickly to its asymptotic value. This is characteristic (see Section 4.4.1) of the SK growth mode and thus agreed with the results of Kolaczkiewicz and Bauer (1986). Comparison between the time dependence of the Gd and the W Auger peaks led to the conclusion that the abrupt change in the Gd intensities, corresponding to the onset of islanding, occurs at the completion of two Gd monolayers (*ie*, a film of thickness equal to one unit cell). The study of the growth of Sm on Mo(110) by Stenborg and Bauer (1987b,c) showed that Sm grows with close–packed (0001) layers parallel to the Mo surface. At room temperature, the growth occurs in the FM mode at least up to 3 ML. The first monolayer of Sm was concluded to be trivalent. For their study of the layer–dependent core level shifts Mårtensson *et al* (1988) grew Yb epitaxially on Mo(110) and showed that the FM growth mode continues up to at least 4 ML of Yb. Adsorption of Tb, Dy and Er on Mo(110) at room temperature by Shakirova *et al* (1992) showed FM growth for the first 2 ML (Dy and Er) or 3 ML (Tb), but with the first monolayer poorly ordered and subsequent layers essentially disordered. Increasing the substrate temperature to 800 K produced ordered growth, with Tb adopting the SK growth mode after reaching a 2 ML coverage (in agreement with the results for Gd/W(110) by Weller and Alvarado 1986a). However, Dy was found to adopt the SK growth mode after only a 1 ML coverage.

The study by Aspelmeier *et al* (1994) of Gd films of a range of different thicknesses grown on W(110) made an important point regarding the annealing of films to improve the crystallographic order — the maximum temperature that can be used before islanding of the film takes place depends strongly on the film thickness. If this is overlooked when preparing rare–earth films, then it can have a dramatic effect on the resultant film structure. AES(*t*) data from Gd films grown with the W substrate at room temperature and at an elevated temperature (Fig. 5.27) showed the characteristics of FM and SK growth modes, respectively. Annealing as–grown films can reduce the density of dislocations and smooth the surface topography, but if the anneal temperature is taken too high then the films can start to island. This is evidenced by an increase in

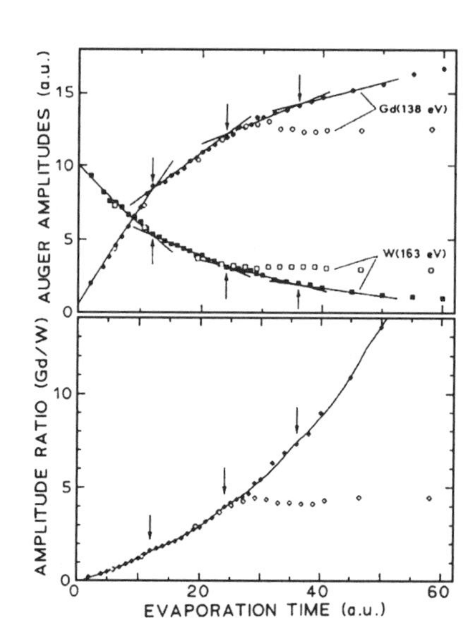

Fig. 5.27 Growth of Gd/W(110) at substrate temperatures of 720 K (open symbols) and 320 K (full symbols). The Auger peak amplitudes of Gd and W (top panel) and their ratio (bottom panel) are characteristic of FM (720 K) and SK (320 K) growth. Arrows mark the completion of the first three layers. From Aspelmeier *et al* (1994).

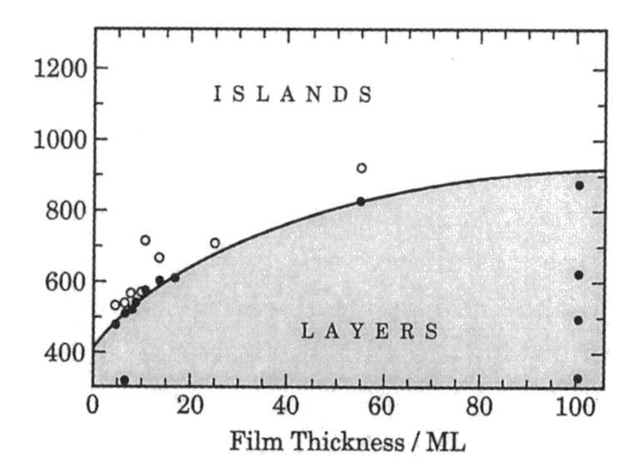

Fig. 5.28 Gd films, grown on W(110) at room temperature, become islanded if they are annealed above a critical temperature which depends on the film thickness. Open symbols are anneal temperatures that are too high, whereas at the full circles a layer–by–layer structure is guaranteed. Optimised anneal temperatures are close to the line. Adapted from Aspelmeier *et al* (1994).

the intensity of the AES signal from the W substrate, penetrating the 1–2 ML coverage of the adsorbate between the thick islands. Aspelmeier *et al* noted that the critical annealing temperature, below which the films remain in a layered structure but above which they start to form islands, is a function of the thickness of the film, varying from 500 K for 5 ML films to 900 K for 100 ML films (Fig. 5.28). The need for careful control of anneal temperatures is clear when it is noted that a change of temperature of only 10 K, especially at low coverages of Gd, can be sufficient to transform a layer–by–layer film into an islanded film.

The AES(t) data for room–temperature deposition of Sc on W(100) taken by Lamouri *et al* (1995) show breaks in the Sc and W Auger peak intensities after the formation of 1 ML and no further break points at higher coverages (Fig. 5.29), indicating a SK growth mode. The open

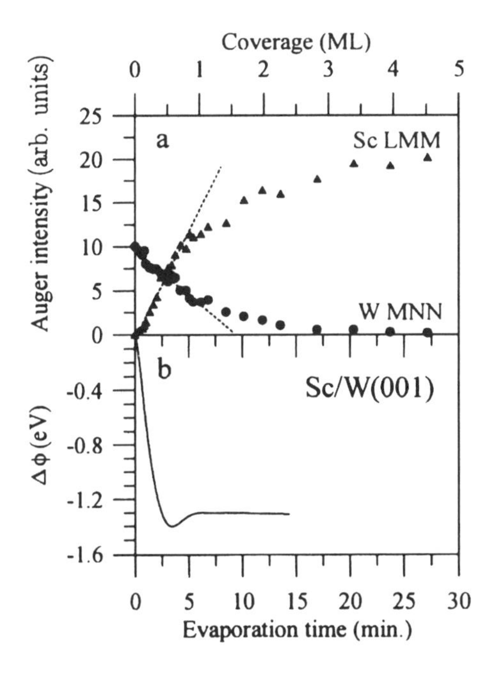

Fig. 5.29 (a) Sc and W Auger peak intensities and (b) the work function of the W(100) surface as a function of Sc coverage. From Lamouri *et al* (1995).

structure of the bcc (100) surface, compared to the relatively close–packed (110), influenced the structures of the islanded films that resulted from high–temperature annealing, but did not affect the basic growth modes.

The growth of Gd/W(111) carried out by Guan *et al* (1995) formed part of a study of the effect of various metallic adsorbates on the W(111) surface. Their aim was to understand the modification of the substrate caused by the growth of ultra–thin metal films, and in particular the surface reconstruction into pyramidal {211} facets observed for Pt, Pd and Au overlayers (see references in Guan *et al* 1995). They concluded that the Gd adopts the SK growth mode and that there is no faceting of the W(111) substrate over the range of coverages of 0.5–5 ML and annealing over the range 300–1500 K.

AES data from the growth of rare–earth metals on transition–metal substrates must be treated with some caution as the possibility of intermixing between the adsorbate and substrate atoms can change significantly the intensities of the Auger signals. Bertran *et al* (1991) observed clear breaks in their AES(t) data from the room–temperature growth of Eu/Pd(111) (Fig. 5.30). From the AES(t) and RHEED data, they

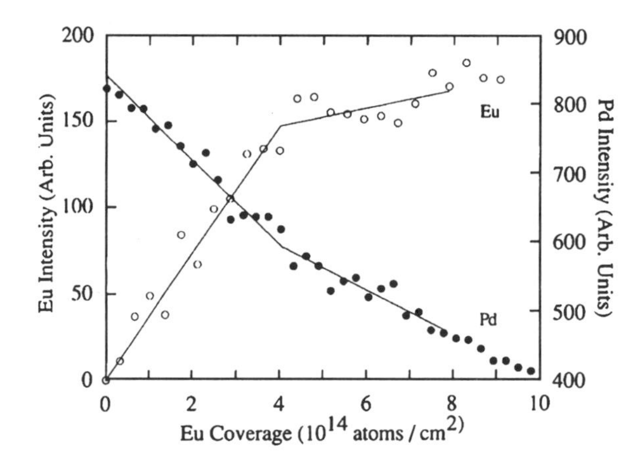

Fig. 5.30 Eu and Pd Auger intensities as a function of Eu coverage. The breaks in the curves correspond to a (2×2) monolayer of Eu. From Bertran *et al* (1991).

deduced a FM growth mode up to 2 ML in which the Eu atoms adopt a (2×2) structure, which becomes unstable as the third monolayer develops, presumably due to interdiffusion between the Eu and Pd. In a study of Sm/Ni(111) Roe *et al* (1994) observed a well–defined break (corresponding to a two–monolayer coverage) in the curve of the Sm Auger signal with

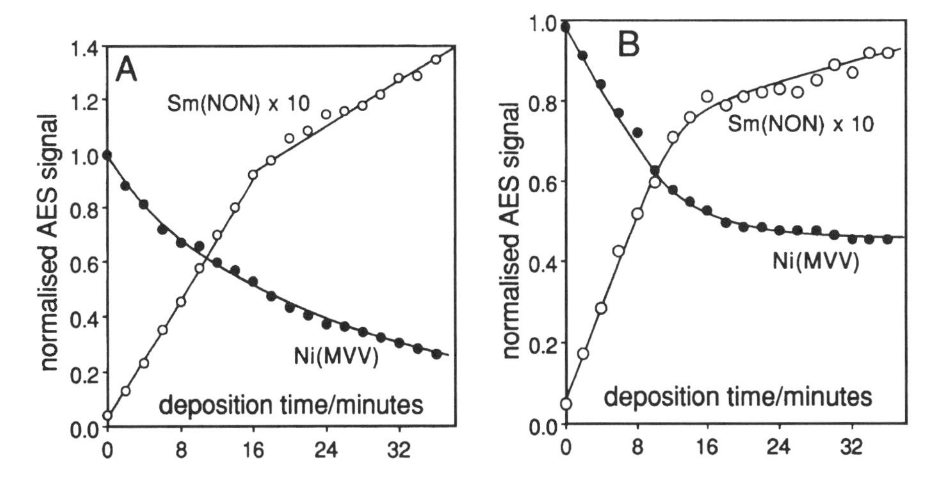

Fig. 5.31 Sm and Ni Auger intensities as a function of Sm deposition time with the substrate at (a) room temperature and (b) 800 K. Adapted from Roe *et al* (1994).

time for room–temperature deposition (Fig. 5.31a) and deduced a SK growth mode. Growth at 800 K produced an AES(t) curve with changes of gradient that were less well defined (Fig. 5.31b) and followed the characteristic shape expected for a SM growth mode. The study of Sm on Co(0001) by Gourieux *et al* (1996) was used in Section 4.4.1 (Fig. 4.6) to illustrate the use of AES(t) to distinguish different growth modes —in this case a model of SM growth fitted the AES(t) data significantly better than a model of FM growth. Ce overlayers grown at room temperature on Pt(111) by Baddeley *et al* (1997) produced AES(t) curves that showed a distinct break point that was taken to define a coverage of one monolayer, and a general shape consistent with the SK growth mode beyond 1 ML. Annealing coverages of greater than 1 ML above 650 K resulted in ordered surface alloys of essentially constant composition. Deposition of Ce at a substrate temperature of 900 K produced a smoothly varying AES(t) curve with no break points, as would be expected with surface alloy formation.

In Section 3.1.3.1 the method of measuring SXRD intensities while scanning along a crystal truncation rod (CTR) in reciprocal space was illustrated with data taken by Nicklin *et al* (1996) during the growth of Sm on Mo(110). The data in Fig. 3.10 were fitted to a theoretical model which provided structural information such as the height of the first Sm layer. In Fig. 5.32 data taken over a longer time interval are shown to illustrate how SXRD has been used to gain information about the growth mode of a rare–earth metal film. The top curve (scanned along the $(00l)$

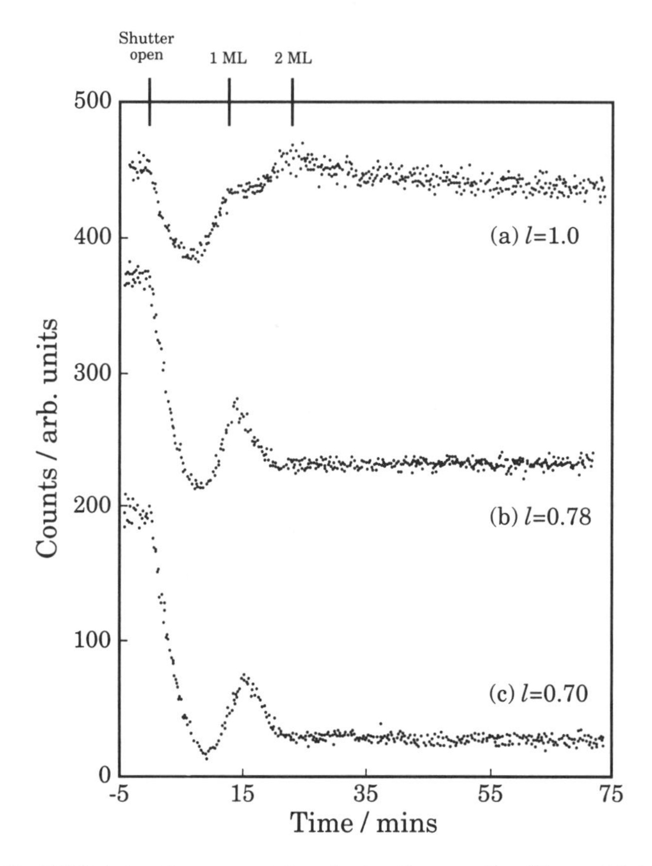

Fig. 5.32 SXRD intensity curves taken during the growth of Sm on Mo(110). The points at which the first two layers are formed are indicated. The time required to complete the second layer is reduced in comparison to the first because of the lower packing density for the divalent hexagonal layer. Adapted from Nicklin *et al* (1996).

rod, where $l = 1$) has maximum sensitivity to the nucleation of islands with perpendicular lattice spacing equal to the bulk Mo(110) layer separation of 0.223 nm. The two lower curves have l values that were chosen to give maximum sensitivity to trivalent ($l = 0.78$) and divalent ($l = 0.70$) Sm atoms at the bridge site, with predicted heights of 0.286 nm and 0.316 nm, respectively. The curve for $l = 1$ shows a shoulder at the completion of the first monolayer and a maximum at the second. The other two curves show maxima at the completion of the first monolayer, but after the following minima the intensities remain low. The data are consistent with a change from FM to SM growth (layer–by–layer to rough growth) shortly before the completion of the second monolayer. It was suggested that this behaviour was a consequence of the 15% different in lattice constants for divalent and trivalent Sm (see Section 5.2.1.2).

5.3 Electronic Structure

Studies of the electronic structure of rare–earth metal surfaces have shown distinct differences between the surfaces of bulk single–crystal samples and those of epitaxial thin films. Some of these differences can be related to the different levels of surface contamination that exist due to the different preparation techniques (covered in Chapter 4) and others to the degree of surface crystallographic order. The principal technique that has been used to probe the electronic structure of rare–earth metal surfaces is photoelectron spectroscopy (also called photoemission), and so this technique will be described before we cover some of the studies that have provided information about the surface geometric structure.

5.3.1 Photoelectron Spectroscopy

Photoelectron spectroscopy has been one of the most extensively used probes of rare–earth metal surfaces to date. X-ray photoelectron spectroscopy (XPS) was used throughout the 1960s and 1970s to study the densities of states of most of the lanthanides and many of their compounds in an attempt to understand the nature of the $4f$ core levels (much of this early work has been reviewed by Baer and Schneider 1987). In the 1970s and 1980s lower–energy photons were used for ultraviolet photoelectron spectroscopy (UPS) studies of the valence–band states, concentrating on the sd–bands within ~ 10 eV of the Fermi level rather than the $4f$ core levels. The lower kinetic energy of the photoelectrons in UPS, of the order of tens of eV rather than hundreds, results in shorter inelastic mean free paths and hence greater surface sensitivity than for XPS. For this reason, the early studies of the surface electronic structure of rare–earth metals were based primarily on UPS, although other low–energy electron spectroscopies, such as electron energy–loss spectroscopy (EELS), also contributed. Angle–resolved UPS (ARUPS) experiments measured the momentum components of the photoelectrons as well as their energies, providing detail of the electronic band structure of the surface region (reviewed by Barrett 1992a).

The distinction between XPS and UPS was a natural one when the sources of radiation were limited to either laboratory–based x-ray generators (producing K_α emission lines of $h\nu > 1000\,eV$) or inert gas discharge sources ($h\nu < 50\,eV$). With the availability of synchrotron radiation sources that can provide high fluxes of photons right across the energy range 10–1000 eV, the distinction has become rather meaningless. If a study of a particular electron level or band is required, then the photon energy can usually be chosen such that the photoelectrons have kinetic energies in the range of a few tens of eV, producing the desired surface sensitivity. Many have adopted the term 'soft x-ray' to denote the photon energy range of 100–1000 eV, but the acronyms XPS and UPS are so deeply embedded in the literature that they will continue to be used for some time to come.

Using the diffraction of the outgoing photoelectron as a probe of local structure is a relatively recent development that has yet to find widespread application for the rare–earth metals (the exception being the study described in Section 5.2.4.1). Photoelectron diffraction aside, photoelectron spectra can still give a great deal of information about the geometric structure of the surface simply because the electronic and geometric structures are so intimately inter-related. The following sections discuss the effects that geometric structure can have on electronic structure and how they are manifested in photoelectron spectra.

5.3.2 Core–Level Shifts

Applying the conservation of energy to an atom in a surface before and after a photoemission event leads to the well–known result that the photoelectron must have a kinetic energy given by

$$\mathrm{KE} = h\nu - E_b - \Phi \qquad (5.1)$$

where E_b is the binding energy of the electron and Φ is the workfunction of the surface. If all of the atoms being probed in an XPS experiment have identical electronic environments, then the binding energies of all of the electrons in a particular core level will be the same and the XPS spectrum will show a single peak corresponding to this binding energy. (Here, we are not concerned with the possibility of the photoelectrons losing kinetic energy by exciting other electrons in the atoms, leading to 'shake–up' and 'shake–off' satellites in the spectrum (see, for example, Woodruff and Delchar 1994). However, if the local electronic environments of some atoms are different from those of others, then the electron binding energies will be shifted relative to each other because the potential seen by an electron is affected by all the other electrons around it. The differing environments could be the consequence of the neighbouring atoms differing in number, relative positions, species, or some combination of all of these factors, and the binding energy shifts are referred to generically as chemical shifts. In principle, each different environment will give rise to a distinct electron binding energy and hence an XPS spectrum will show a peak at each of the corresponding kinetic energies. In practice, the spectrum will show a peak whose lineshape is the sum of all of these contributions, convoluted with the finite energy resolution of the x-ray source and the electron energy analyser of the XPS system.

If the surface of a single–element material has the ideal bulk–terminated structure then the only chemical shift will be due to the atoms at the surface having a reduced coordination relative to those in the bulk. If the surface is well–ordered, then there will be only one (or perhaps two, depending on the surface crystallography) shifted components in addition to the bulk component. The existence of other components indicates either (i) that the surface is reconstructed or not well–ordered, giving rise to atoms having various coordination numbers, or (ii) the presence of other species of atoms (*ie*, surface contamination). If the photoemission lineshape of a core level from the substrate is monitored as an adsorbate overlayer is formed, then the origin of each component can be identified by decomposing the lineshape at various coverages of the adsorbate, noting the changing intensities of each of the components as the adsorbate atoms influence the substrate atom binding energies (as shown in Fig. 5.33 for Gd/W(110) and used by White *et al* 1995 and Tucker *et al* 1998).

Equation (5.1) assumes that the only change experienced by the atom is the loss of the photoelectron, and that all of the other electrons remain unaffected by the process. In practice, this is never realised because the

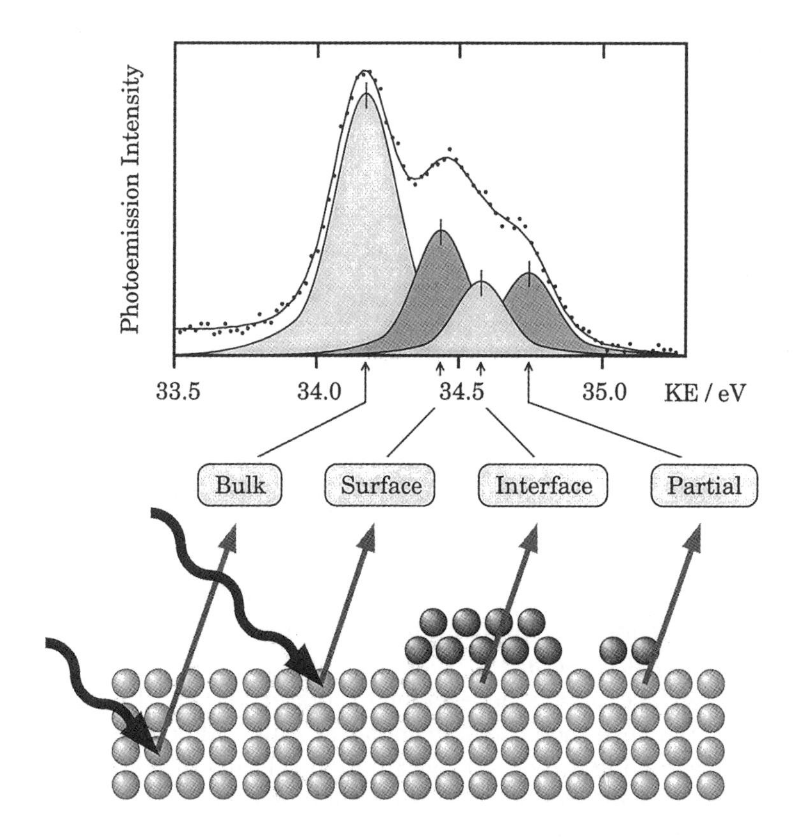

Fig. 5.33 The core–level photoemission lineshape from a substrate that is partially covered by atoms of a different species can be decomposed into a number of distinct components, each corresponding to atoms in a different environment. The spectrum shown is from one of the W 4*f* levels during the growth of Gd on W(110).

atom is left in an excited state and so the remaining atomic electrons will relax to lower energies to screen the core hole left by the departing photoelectron. In addition, in metallic solids the itinerant valence electrons will also make a significant contribution to the screening of the core hole. Both of these contributions to the relaxation, which can be labelled intra-atomic and interatomic, provide additional kinetic energy to the photoelectron and so reduce the binding energy as deduced from equation (5.1). This apparent shift in binding energy, as determined by measurement of photoelectron kinetic energies, can be described as a final–state effect as it is an artefact of the photoemission process rather than a measure of the true chemical shift, which is an initial–state effect. Separating initial–state from final–state effects is not always straightforward, and in many systems the resultant shifts are comparable in magnitude. However, this is not a serious problem as both shifts are dependent on the local electronic environment of the photoelectron, and so this means that the total shift seen in photoemission spectra is characteristic of a particular local atomic structure, regardless of the details of its origin. As the shifts manifest themselves most clearly in photoemission from core levels, they are known as core–level shifts (CLS). By treating the observed CLS as a 'fingerprint' of the local atomic structure, the technique of CLS spectroscopy (known in its earlier days as ESCA — electron spectroscopy for chemical analysis) has found widespread use since the 1960s. The CLS of atoms at a clean surface, typically a few tenths of an eV for the rare–earth metals, is labelled a surface core–level shift (SCLS) to distinguish it from the shifts produced by the proximity of other species of atoms. The SCLS component of a core–level photoemission lineshape can be identified by one of two methods. The surface sensitivity can be changed by increasing the photoelectron emission angle relative to the surface normal — at off–normal emission the effective depth from which the photoelectron can emerge is less than for normal emission and thus the surface component is more intense relative to the bulk component. Alternatively, if the photon energy can be adjusted then the mean–free–path of the photoelectrons (and hence the surface sensitivity) can be varied.

To be able to decompose a lineshape into contributions from electrons in multiple environments requires an electron energy level that has an intrinsically narrow width in energy and an experimental energy resolution that is rather better than the magnitude of the CLS to be studied (which may be from ~ 0.1 eV to many eV). By choosing a suitable core level, preferably one that does not have a complex multiplet structure, and by using a synchrotron to provide a narrow spread of photon energies, the technique of CLS spectroscopy can provide useful structural information on clean surfaces and the adsorption of atoms on a surface or the growth of thin films, including metal–on–metal growth.

A good example of the use of CLS spectroscopy to indicate the surface morphology of thin–film growth is shown in Fig. 5.34. The photoemission lineshape of one component of the Tb 4*f* core–level multiplet is shown for a

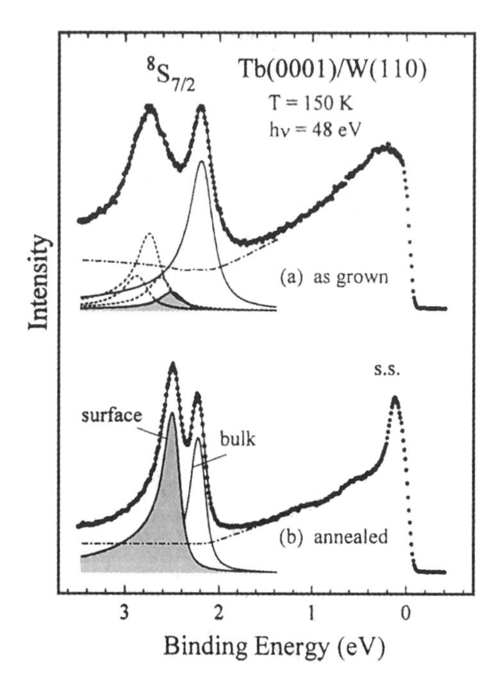

Fig. 5.34 Photoemission spectra from a thin film of Tb grown on W(110) at 150 K. The Tb 4*f* core–level lineshape is decomposed using a least–squares–fit analysis of (a) the as–grown Tb film and (b) the well–annealed Tb(0001) film. Spectrum (b) is described well using only two components, whereas spectrum (a) requires four components to obtain a good fit. Adapted from Navas *et al* (1993).

film at 150 K and for the same film after annealing at 900 K. The spectrum from the annealed film can be fitted to two components; one from bulk atoms and a single contribution from the surface atoms. By contrast, the spectrum from the as–grown film requires four surface components, corresponding to Tb atoms with six, seven, eight and nine neighbouring atoms. In an atomically flat hcp (0001) surface, all the surface atoms have nine neighbours (six in the surface plane and three below) and thus the presence of atoms with less than nine neighbours indicates a rough surface. Thus, the two spectra show clearly the smoothing effect of the 900 K anneal.

5.3.3 Valence–Band Structure

The contribution of electron spectroscopy to the study of the geometric structure of surfaces has been dominated by CLS and SCLS spectroscopy. Studies of the valence–band electronic structure give geometric information in a much less direct way and have been limited to observations of surface states that are sensitive to the degree of surface crystallographic ordering. Surface states are electronic states that are localised to the surface region,

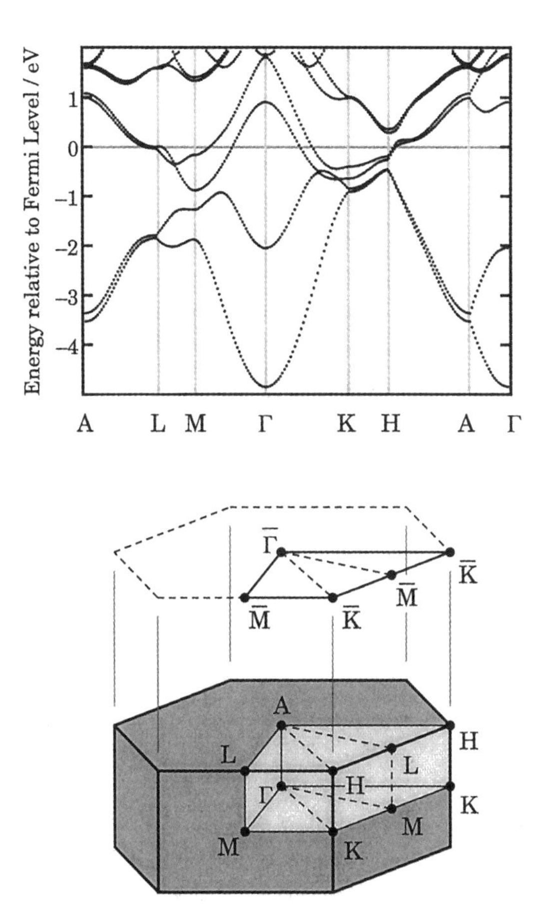

Fig. 5.35 The bulk band structure of the rare–earth metals (without the $4f$ electron levels) and the Brillouin zone for the hcp crystal structure. The projection of the Brillouin zone onto the plane perpendicular to ΓA defines the surface Brillouin zone for a (0001) surface.

having either (i) energies that lie in the energy gaps of the bulk band structure, or (ii) symmetries different from those of the bulk states at the same energy. For the rare–earth metals that adopt the hcp crystal structure, the generic band structure and Brillouin Zone are shown in Fig. 5.35 (this is actually the band structure for Y—for the lanthanides the band structure will appear similar except for a slight expansion of the energy scale and the addition of the highly localised $4f$ states). The [0001] direction corresponds to the direction labelled ΓA in reciprocal space, and it can be seen that for this direction there is a band gap around the Fermi energy.

Metal surfaces can produce two types of surface states—Tamm states and Shockley states. Tamm states are not often observed as distinct features in UPS valence–band spectra due to their proximity in energy to other (bulk) states. Shockley states are solutions of the Schrödinger equation that take account of the potential at the surface, and often manifest themselves in valence–band spectra as sharp features just below the Fermi energy. The existence of such a state is taken to indicate a clean and well–ordered surface, as there is substantial evidence that disruption of the surface through roughening or the addition of even small concentrations of surface contaminants can produce significant attenuation of these states.

Valence–band photoemission spectra from the surfaces of many bulk single crystals of the rare–earth metals have exhibited a feature that has been shown to be dependent on the degree of surface ordering (Barrett 1992a), but does not exhibit the characteristics of either a Tamm state or a Shockley state. Although this surface–order–dependent state was used to indicate the extent of the surface reconstructions of the $(11\bar{2}0)$ surfaces of Y (Barrett *et al* 1987b, 1991b) and Ho (Barrett *et al* 1991a) described in Section 5.2.1.1, its origin has yet to be determined.

Further Reading

Many of the earlier results of photoemission experiments on the surfaces of bulk single–crystal rare–earth metals have been reviewed by Barrett (1992a) and experiments on epitaxial thin films have been reviewed by Gasgnier (1995).

5.4 Surface Magnetism

The surface magnetism of the rare–earth metals is an expanding field of research which is both technologically important and scientifically challenging. The drive to study the magnetism of the rare–earth metals and rare–earth–based systems has its origin in the desire to understand those magnetic properties that can be exploited in magnetic recording technology and sensor applications. Such properties include perpendicular magnetic anisotropy, oscillatory exchange coupling and giant magnetoresistance. Understanding the physical mechanisms that underpin these properties remains a formidable challenge, but is a prerequisite to a development of future applications.

Compared to the studies of geometric and electronic structure, work on the magnetic structure of rare–earth metal surfaces has seen a recent revival of interest. This is in part due to the application of techniques such as magnetic dichroism in surface–sensitive spectroscopies throughout the 1990s (see, for example, van der Laan *et al* 1996) and more recently scanning tunnelling spectroscopy (see, for example, Bode *et al* 1998a).

To a first approximation, the highly–localised $4f$ electron orbitals are responsible for the magnetic moments of rare–earth atoms. Thus it would be expected that the rare–earth metals are archetypal local–moment ferromagnets whose properties can be readily understood and predicted. This may be the case for the bulk magnetic properties, but the surface magnetism is another matter. Various surface phenomena have been observed that complicate the otherwise straightforward picture of bulk magnetism. An example is surface–enhanced magnetic order (SEMO), which can give rise to a situation where, at temperatures above the ordering temperature of the bulk, a magnetically ordered surface exists on a magnetically disordered paramagnetic bulk material (Fig. 5.36). Another aspect of interest is the magnetic coupling between the moments of the atoms in the surface layer and those of the bulk atoms below. If the ferromagnetic ordering of a bulk ferromagnet continues to the surface, then the magnetic moments of surface atoms will be parallel to those in the bulk. However, a 'surface magnetic reconstruction' may occur if the surface spins have a different alignment, such as antiparallel to the bulk spins if the surface–to–bulk coupling is antiferromagnetic.

Attempts are being made to understand such phenomena, but one of the impediments to progress is the inconsistent, sometimes contradictory,

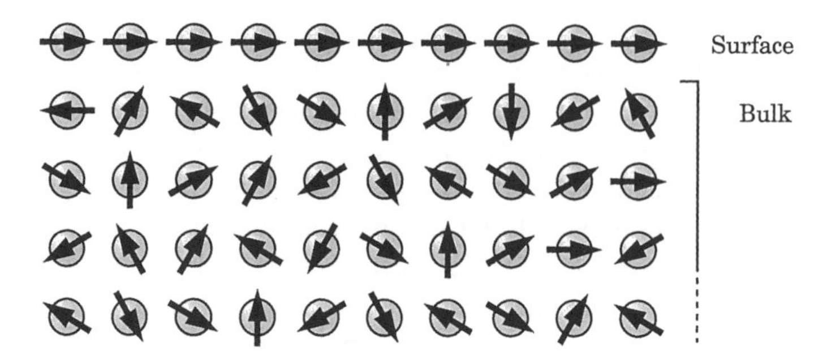

Fig. 5.36 The orientation of magnetic moments for a ferromagnetic material exhibiting surface–enhanced magnetic ordering (SEMO). At temperatures above the bulk Curie temperature the surface can be ordered even when the bulk is disordered.

results produced by various research groups. Even for Gd, the most studied of all the rare–earth metals, there is still disagreement over the magnitude of the SEMO enhancement, ranging from a few kelvin to many tens of kelvin. Also, contradictory evidence (both experimental and theoretical) exists for surface–to–bulk coupling being ferromagnetic, antiferromagnetic and imperfect ferromagnetic (which produces surface spins that are canted, neither parallel nor antiparallel to those in the bulk — see Fig. 5.37). All of the studies involved the growth of thin films of the rare–earth metals due to the problems associated with contamination of the surfaces of bulk single crystals — it is known that even small concentrations of contaminants can influence the surface magnetic structure. Also, the morphology of the films grown will influence their magnetic properties, and thus comparing and correlating results from different films, let alone experiments carried out by different research groups, can be meaningful only if the films are completely characterised in terms of their structural and chemical imperfections. This was not always possible if the relevant surface techniques were not available when the magnetic measurements were made, or if it was not considered necessary to characterise the films beyond an estimate of the average film thickness. In some cases the difficulties encountered in preparing clean, crystallographically well–ordered, atomically flat thin films may have been underestimated and so the results of the magnetic measurements compromised.

5.4.1 Surface–Enhanced Magnetic Order

The first observation of SEMO at the surface of Gd was carried out by Rau (1983) using electron capture spectroscopy (the technique is reviewed by Rau (1982a)). This gave a surface Curie temperature of ~ 310 K (*cf* the bulk Curie temperature of 293 K, indicating an enhancement of ~ 15 K). Later experiments involving spin–polarised LEED and the magneto-optic Kerr effect (MOKE) determined an enhancement of 22 K (Weller *et al* 1985a,c,d). More recent spin–polarised photoelectron and secondary electron spectroscopy experiments have confirmed the SEMO effect for Gd (and Tb), with values of the enhancement as large as 60 K (Tang H *et al* 1993a,d) or possibly even 80 K (Vescovo *et al* 1993b). The largest effect reported to date is that derived from a spin–polarised photoelectron diffraction (SPPD) study by Tober *et al* (1998), which indicated an enhancement of 85 K. The steady increase in the magnitude of the apparent enhancement has been attributed to improvements in the quality of the epitaxial thin films grown, but the correlation between SEMO and the structural quality of the films has not yet been established quantitatively.

The origin of SEMO must lie in the surface and bulk electronic structure of the rare–earth metals, as it is the valence electrons that mediate the exchange interaction between the (principally 4*f*) magnetic moments, but the details of the mechanism fall outside the scope of this book. More than fifteen years after the original work of Rau *et al* and Weller *et al*, the enhancement of the surface Curie temperature still attracts a great deal of attention and controversy — throughout the 1980s and 1990s the measured extent of surface enhancement was increasing, but the most recent results based on spin–polarised secondary electron emission spectroscopy (SPSEES) and MOKE indicate that the surface and bulk Curie temperatures are essentially the same (Arnold and Pappas 2000), bringing into question not only the magnitude of the effect, but the very existence of the effect itself. Clearly, SEMO will continue to be a topic of interest for many years to come.

5.4.2 Surface–To–Bulk Coupling

Some of the early experiments observing SEMO on the (0001) surface of Gd also involved techniques that gave information about the surface–

to–bulk magnetic coupling. For thin films of Gd(0001), the magnetisation is in the basal plane of the hcp crystal structure and is thus in–plane.

The spin–polarised UPS (SPUPS) data of Weller *et al* (1985a) indicated an antiferromagnetic surface–to–bulk coupling, which was later supported by electronic structure calculations (Wu RQ *et al* 1991a, Wu and Freeman 1991b, Anisimov *et al* 1994, Heinemann and Temmerman 1994). However, subsequent results from SPUPS (by various groups, including those of Dowben, Erskine and Hopster) and magnetic dichroism (by Kaindl's group) indicated ferromagnetic coupling (in the latter case, for Tb as well as for Gd). Again, the experimental results were supported by electronic structure calculations (Bylander and Kleinman 1994, Eriksson *et al* 1995) based on the local–spin–density approximation (LSDA). By measuring both the in–plane and out–of–plane components of the electron polarisation in SPUPS and spin–polarised secondary electron emission spectroscopy (SPSEES), Tang H *et al* (1993a,d) showed that the surface magnetic moments are canted, producing a measurable out–of–plane component to the surface moment (Fig. 5.37).

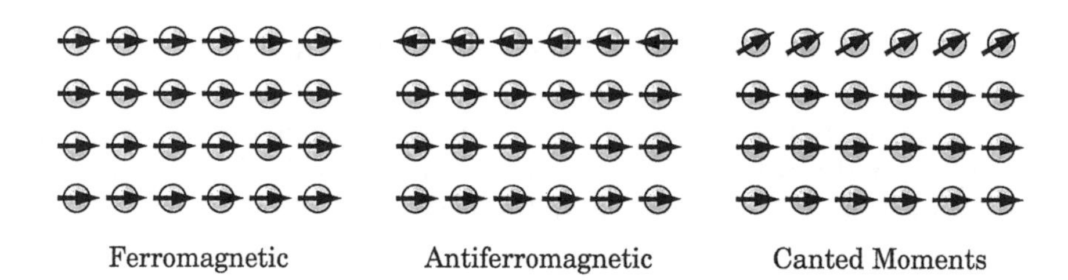

Fig. 5.37 Proposed models for the surface–to–bulk magnetic coupling in rare–earth metals (the top layer of magnetic moments represents the surface layer). The first experimental evidence indicated antiferromagnetic coupling, later experiments preferred ferromagnetic coupling, and the most recent experiments indicate that the surface moments are canted out of the surface plane.

The surface–to–bulk coupling of magnetic moments will be discussed further in Chapter 7, where the link between the magnetic and geometric structure of surfaces is discussed in the context of surface calculations.

Further Reading

Surface magnetism of the rare–earth metals, including SEMO and surface–to–bulk coupling, has been reviewed by Dowben *et al* (1997). An overview of the theory underlying surface magnetic reconstructions has been given by Rettori *et al* (1995). An earlier, and more general, review of surface magnetism was carried out by Siegmann (1992).

CHAPTER 6

QUANTITATIVE LOW–ENERGY ELECTRON DIFFRACTION

The basic principles of LEED, as used for a qualitative understanding of surface atomic structure, have already been outlined in Chapter 3. In this chapter, we adopt a more quantitative approach to LEED which will enable us to determine surface atomic positions with a high degree of accuracy. The relative dimensions of the surface unit cell can be determined solely from the positions of the diffraction beams in a LEED pattern. However, atomic positions can be obtained only using the variation of the diffraction beam intensities as a function of incident electron beam energy or angle. Since the incident electron beam energy is controlled by the accelerating voltage (V) of the electron gun, these intensity (I) spectra are known as LEED I–V spectra or curves.

We start by outlining one approach which may be used to acquire LEED I–V spectra and then describe how theoretical calculations are used to determine atomic positions. Structure determinations using LEED proceed in an iterative manner, in which the correspondence between theoretical and experimental I–V curves are quantified using reliability factors, or r-factors. The input parameters of the calculations are adjusted, new r-factors are calculated and the whole process is repeated until the lowest possible r-factor is obtained. Theoretical calculations need to model the strong scattering of low–energy electrons within a solid because it is this that gives LEED its surface sensitivity. The inability to calculate this scattering realistically was, in part, responsible for the five–decade hiatus between the discovery of electron diffraction (Davisson and Germer 1927) and the first structural determination using LEED. There is more than one way to solve the LEED problem and the aim here is to give an idea of how I–V spectra are calculated rather than a rigorous mathematical treatment.

6.1 Quantitative LEED experiments

Quantitative LEED can involve measuring either the intensity profile of an electron diffraction beam (spot profile analysis, or SPA-LEED) or the variation of its integrated intensity as a function of incident electron beam energy to produce an *I–V* spectrum. The former can be used to study defects at surfaces and to investigate thin–film growth modes (Zuo *et al* 1994), whereas the latter is used to determine surface atomic positions. There are several approaches to recording *I–V* spectra, with the choice dependent on the system under study. If the system is susceptible to electron beam damage or stimulated desorption, a very low intensity of incident electrons is required which in turn results in a weak intensity diffraction pattern. In such cases the fluorescent screen in standard LEED optics (Fig. 3.1) is not capable of producing an observable image, so that either an electron multiplier has be used or the screen has to be replaced entirely by a two–dimensional detector. However, for many surface studies a standard set of commercially available LEED optics, together with a data acquisition system capable of digitising the LEED pattern, is often all that is required to record *I–V* spectra. In a typical quantitative LEED experiment the incident beam of electrons impinges on the sample normal to the surface at an energy set by the controlling computer, the diffraction

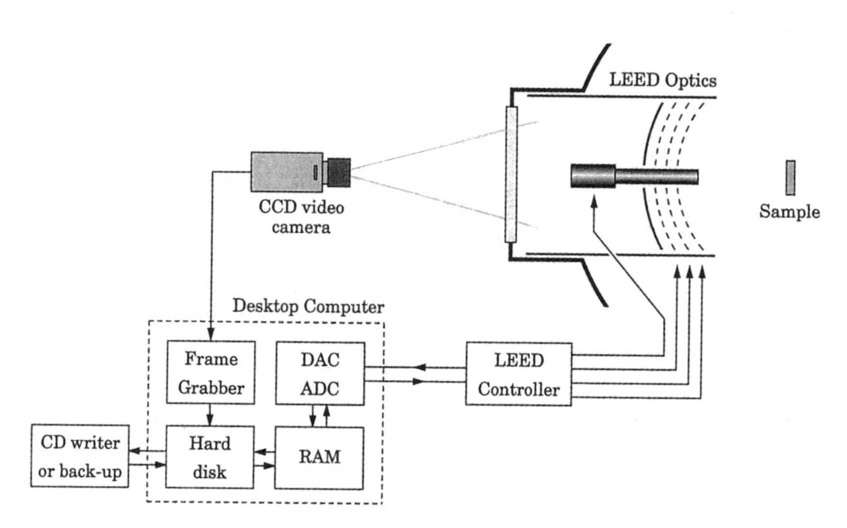

Fig. 6.1 Typical video–based system for the acquisition of digital images of LEED patterns, used with conventional LEED optics.

pattern is digitised and the background–subtracted intensities of all the visible diffraction beams are extracted and stored (Fig. 6.1). This routine is repeated over a wide energy range to produce *I–V* spectra for as many diffraction beams as possible. The *I–V* spectra of diffraction beams that are equivalent by symmetry are compared with each other using r-factors to ensure accurate sample alignment — if the incident beam is off–normal then diffraction beams that should be equivalent at normal incidence will show intensity variations due to the broken symmetry. The final set of *I–V* spectra to be used in the structure determination are calculated as the average of the *I–V* spectra from all symmetrically equivalent beams. This whole procedure can be repeated with the electron beam incident at a small angle (~ 10°) to the surface normal, since the broken symmetry enables the collection of more *I–V* spectra from the same set of diffraction beams. As the *I–V* spectra are very sensitive to the size of the off–normal angle, this needs to be known accurately. In such a geometry the specularly reflected (00) beam, which is not visible for the normal incidence geometry, also becomes visible.

6.1.1 Electron scattering

We have already noted in Chapter 3 that LEED owes its surface sensitivity to the strong elastic and inelastic scattering that low–energy electrons suffer within a solid. If a monoenergetic beam of electrons is incident on a surface then the electrons will undergo forward and backward scattering, with the amount in each direction dependent on the kinetic energy of the electrons. Those electrons that suffer large energy losses due to inelastic collisions, plasmons and valence–band transitions are rejected by the retarding grids of the LEED optics (Fig. 3.1). The observed LEED pattern therefore comprises only the elastically backscattered electrons plus the electrons that have sustained only minor alterations to their energy due to the creation or annihilation of phonons within the solid. The number of these quasi-elastic electrons that have suffered phonon interactions is difficult to determine, but it is generally considered appropriate to consider such electrons as inelastic in the theory even though experimentally they contribute to the measured intensity of a diffraction beam. All this inelastic scattering is represented in the theory of LEED by introducing a single parameter, known as the mean free path, which effectively decreases the contribution from elastically scattered electrons. This may seem a crude approximation, but we can justify this by considering that we are not concerned with how the electrons lose their energy, but only with how many do so. The mean free path is incorporated into a LEED calculation via a single parameter, which will be covered in Section 6.2.1.

6.1.2 Structure and LEED I–V

In the case of x-ray diffraction the interaction of the photons with the crystal is weak, so that the photon is assumed to scatter elastically once, at most, before leaving the crystal and reaching the detector. A kinematic analysis of the diffraction intensities can be applied and structural information can be obtained after relatively modest mathematical manipulation of the experimental diffraction spectra. However, low–energy electrons in a crystal may undergo elastic scattering many times before escaping and contributing to the intensity of a diffraction beam. This means that a kinematic approach is not suitable for calculating diffraction intensities of low–energy electrons. However, a kinematic approach is perfectly valid for determining the positions of the diffraction beams since they are independent of the complexity of the process of electron scattering. The *I–V* spectra cannot be manipulated directly to yield structural information because of multiple scattering, which introduces additional peaks at energies different to those of the Bragg peaks predicted by a kinematic analysis. In addition, the Bragg peaks themselves can appear at energies lower than those predicted by the kinematic analysis. The multiple scattering of low–energy electrons demands that a structural determination using LEED adopts a more convoluted procedure. This involves postulating a structure, calculating the *I–V* spectra for such a structure, comparing the experimental and theoretical spectra using r-factors and adjusting the parameters of the calculation until the two sets of spectra are as closely matched as possible. Some of the parameters in these calculations are directly related to the geometric structure, but other non-structural parameters also have a significant effect on the final LEED *I–V* spectra. The formalisms used in calculating *I–V* spectra, the parameters involved in such calculations and the methods employed to optimise the correspondence of experimental and theoretical spectra will be the subjects of the remainder of this chapter.

6.2 Dynamical Calculations

A first attempt to interpret experimental *I–V* spectra is to assume that the surface of the system under study is a simple truncation of the bulk structure, assuming that the symmetry of the LEED pattern allows such a proposition. If an adsorbate species is present then a bulk truncation of the substrate may still be assumed with the initial adsorbate location determined by the symmetry of the LEED pattern. It should be noted, however, that several different adsorbate positions can result in the same LEED pattern. Thus, one probable site is chosen in the first instance, but all possible adsorbate sites should be tested. The process of calculating *I–V* spectra for a postulated structure is usually separated into several parts, with the system being modelled by a slab comprising several layers. Firstly, the scattering from a single atom of each atomic species present is calculated with due regard for thermal effects. Next, the multiple scattering within a single layer, which may comprise several different atomic subplanes, is calculated. Finally, the multiple scattering from a collection of such layers, stacked vertically to represent the complete crystal structure, is calculated and the desired *I–V* spectra for each symmetrically inequivalent diffraction beam is generated.

6.2.1 Atomic Scattering

A kinematic analysis of x-ray diffraction assumes that a plane wave incident on an atom is scattered into another plane wave. This is the Born approximation, in which the scattering depends only on the momentum transfer of the photon. The use of such an approximation is justified because the interaction of x-rays with atoms is weak. However, for low–energy electrons much stronger interactions require a full quantum mechanical description of the scattering characteristics of an atom in a solid.

It is tempting to assume that the interaction of an electron with an atom in a solid can be approximated by the scattering due to an isolated atom. However, even though the potential associated with such an atom is easily calculated, it is not suitable since low–energy electrons are sensitive to the bonding or valence electrons in a solid. The scattering which we wish to calculate is, in principle, complicated because it is a many–body problem involving the solution of a Schrödinger equation which describes the interaction of a low–energy electron with all of the charged particles of a solid. The potential we need to calculate therefore includes the electrostatic effects of the ion cores and the valence electrons of the solid as well as the influences of exchange (which takes into account the indistinguishability of electrons) and correlation (which takes into account the fact that the behaviour of one electron is correlated with the behaviour of all other electrons in the system). In reality, a much simpler representation of the atomic scattering in a solid is generated using a hypothetical structure consisting of touching spheres with spherically symmetric potentials, with regions of constant potential in the interstitial volumes. This is the well–known muffin–tin potential model (Fig. 6.2). The constant potential is called the muffin–tin zero and can be thought of as modelling the valence electrons.

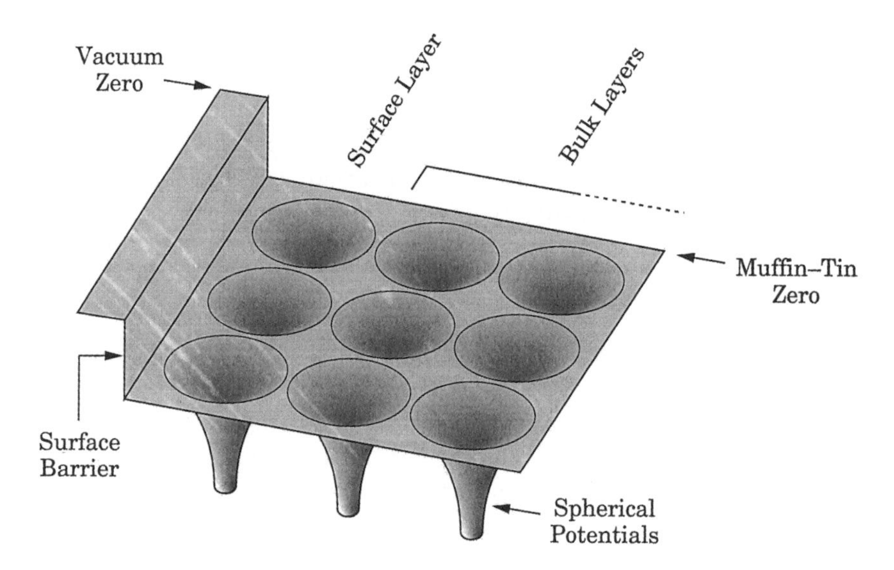

Fig. 6.2 A muffin–tin potential model for a two–dimensional slice through the surface of a crystal. The step potential at the surface, between the potential inside the crystal and the zero potential of the vacuum outside the crystal, is an approximation to the smoothly varying surface barrier.

A scattering potential has to be generated for each different type of atom present in the surface, since each chemical species has different angular and energy–dependent scattering properties. These potentials are usually calculated by incorporating atomic potentials in a cluster calculation that models adsorbate atoms on a surface. Surprisingly, such approximations seem to give reasonable results even though the electronic structure of adsorbate atoms on a surface may not be remotely like those in a cluster, and may not have spherical symmetry. This obviously implies that the LEED intensities are far more sensitive to atomic positions than to the fine detail of the scattering potential.

Having defined a model structure, we are now in a position to calculate the scattering properties of the material. We should note that if we have an incomplete outer shell then the spherical symmetry inside the muffin–tin sphere will not be conserved. To regain spherical symmetry we force each outer–shell electron orbital to be occupied equally by the total charge in the outer shell divided by the number of orbitals in the outer shell. One further point is that if we assume free–atom wavefunctions inside the muffin–tin spheres then some of these wavefunctions may extend past the muffin–tin sphere radius. If this is the case then they have to be truncated, which requires additional corrections.

We now need to solve the Schrödinger equation using a total wavefunction for all of the core–state electrons plus the incident low–energy electron. Since the core states are not easily polarised by the incident electron, the complicated total wavefunction of the system can be represented as a product of separate core–state wavefunctions, $\psi_j(\boldsymbol{r}_j)$, and the wavefunction of the incident electron, $\phi(\boldsymbol{r}_0)$. The resulting wavefunction must be antisymmetric under exchange of particles because we are dealing with electrons, which are fermions. Neglecting spin, the many–body Schrödinger equation can then be reduced to a one–electron equation. A spherically symmetric potential implies that $\phi(\boldsymbol{r})$ inside the muffin–tin sphere can be decomposed into a radial component, $R_l(r)$, and spherical harmonics, $Y_{lm}(\theta,\varphi)$, so that the Schrödinger equation may be expressed in spherical coordinates (see, for instance, Mandl 1992)

$$\left[-\frac{\hbar^2}{2mr^2}\frac{\mathrm{d}}{\mathrm{d}r}\left(r^2\frac{\mathrm{d}}{\mathrm{d}r}\right)+\frac{\hbar^2 l(l+1)}{2mr^2}+V_n+V_{ee}+V_{ex}\right]R_l(r) = E\,R_l(r) \qquad (6.1)$$

where V_n is the screened potential of the nucleus, V_{ee} is the electron–electron Coulomb repulsion and V_{ex} is the exchange–correlation potential. Using the Slater approximation, V_{ex} may be set to be proportional to the square root of the charge density $\rho(\boldsymbol{r})$ of the core–state electrons. Equation

(6.1) is the one that is generally used to calculate ion–core scattering. We first determine the solutions of this equation within the muffin–tin sphere by numerical integration and then match these results to solutions outside the sphere. Where the muffin–tin potential is constant, the potential within the muffin–tin spheres has no electrostatic influence on a low–energy electron and the exchange term is zero, so equation (6.1) simplifies to

$$\left[-\frac{\hbar^2}{2mr^2}\frac{\mathrm{d}}{\mathrm{d}r}\left(r^2\frac{\mathrm{d}}{\mathrm{d}r}\right)+\frac{\hbar^2 l(l+1)}{2mr^2}\right]R_l(r) = E\,R_l(r) \tag{6.2}$$

Since the ion–core potential is current conserving, the amplitudes of the incoming and outgoing waves must be equal, which means that the incoming and outgoing waves can only be shifted in phase. The solution of equation (6.2) therefore comprises an incoming and outgoing wave with the difference between the two expressed as a phase shift (δ_l)

$$R_l(r) = \beta_l\left[\,\exp(2i\delta_l)\,h_l^1 + h_l^2\,\right] \tag{6.3}$$

where β_l is the amplitude of the incoming wave with angular momentum l, δ_l is the phase shift for a spherical wave with angular momentum l and h_l^1 and h_l^2 are spherical Hankel functions. The incoming and outgoing waves have the same angular momentum, since a spherically symmetric potential cannot change this quantity. Once we have determined the solutions inside and outside the muffin–tin sphere, the solutions are matched at the boundary using a logarithmic derivative.

Thus the scattering by a muffin–tin sphere or ion core can be represented by a set of energy–dependent phase shifts (one for each angular momentum quantum number) and once these are known we can determine the scattering by an ion core for any incident wave. Therefore, the wavefunctions inside the muffin–tin sphere do not need to be determined for each LEED calculation. With the phase shifts in hand we can show that, for a plane wave incident on an ion core, the scattered wave at large distances from the core is given by

$$R_l^{(s)}(r) \underset{r \to \infty}{=} t(E,\theta)\left[-\frac{\exp(ikr)}{r}\right] \tag{6.4}$$

where $t(E,\theta)$ is known as the t-matrix and describes the angular–dependent scattering amplitudes for a given electron energy such that

$$t(E,\theta) = \frac{1}{(2ik)}\sum_l (2l+1)\left[\exp(2i\delta_l)-1\right] P_l\cos(\theta) \tag{6.5}$$

where P_l are the familiar Legendre polynomials. The sum over l is from zero to infinity, so we need to find a maximum value beyond which the scattering is not altered significantly by the inclusion of extra terms. The maximum value of l depends on the particular structure being determined and the energy range of the experiment. In general, the higher the kinetic energy of the incident electrons and the larger the atomic radii, the more l terms we will have to include in equation (6.5). Fig. 6.3 shows the phase shifts for Sc used in the structure determination of Sc(0001) by Dhesi *et al* (1995). It can be seen that, because the t-matrix elements are periodic in π, we need only express the phase shifts over this interval. We also see that, with increasing energy, more and more phase shifts are required to describe the complicated angular structure of the scattering.

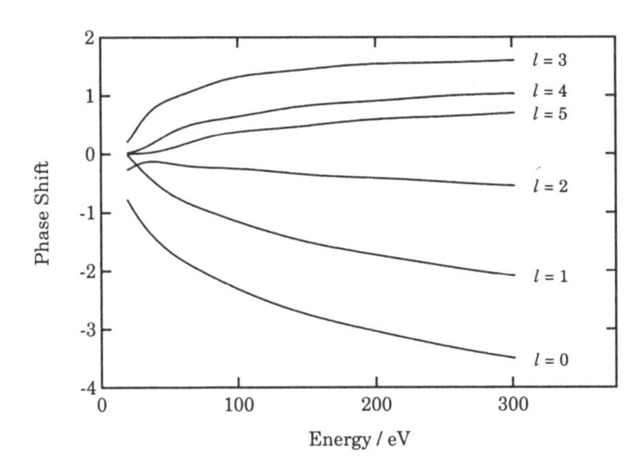

Fig. 6.3 Phase shifts for Sc from $l = 0$ to $l = 5$.

With the scattering by the muffin–tin spheres described completely by a set of phase shifts, it is useful to imagine the spheres in our model as points immersed in a constant potential. The scattering between the points is then described by propagator functions which can be represented in a plane or spherical wave basis (to be covered in more detail in Section 6.2.2). The constant potential has to be referenced to some energy level of the system, which is usually the vacuum level (Fig. 6.2). The value of the constant potential can be calculated by solving the relevant Schrödinger equation, but it is generally adjusted during a structural determination to optimise the fit between experimental and theoretical I–V spectra. The parameter which is adjusted during this optimisation is referred to as the inner potential. However, the inner potential and the muffin–tin zero are not, strictly speaking, identical quantities. The muffin–tin zero is a concept which represents the potential due to the valence electrons in a solid, whereas the inner potential is a measurable quantity related to the acceleration of an electron as it passes through the potential barrier at the vacuum–solid interface. The constant potential also has an imaginary component which represents the inelastic scattering of the waves propagating between the point scatterers. This imaginary component can be calculated, but variations in its value are small from one material to another and for different electron kinetic energies; a value of –4 eV is typical and represents a mean free path of ~ 0.5 nm for an electron with a kinetic energy of 100 eV.

The final point to discuss with regard to atomic scattering is the influence of thermal vibrations on diffraction intensities. In practice, multiple–scattering calculations generate temperature–dependent phase shifts to model the effects of thermal vibrations, which are assumed to be isotropic and therefore angular momentum conserving. However, here we will restrict ourselves to a kinematic approach to understand the physical consequences of thermal vibrations. Experimentally it is found that an increase in the temperature of a surface causes a reduction in the diffraction intensities, an increase in the background intensity between diffraction beams and a broadening of the spots. In addition, the reduction in the diffraction intensities is more marked at higher incident electron energies than at lower. These effects are due to thermal vibrations of the surface atoms inducing more surface disorder. In turn, the resulting destructive interference between the scattered waves from the disordered atoms effectively transfers intensity from the diffraction beams to the background. The phase shifts due to thermal vibrations are larger for backscattering than for forward scattering and increase for smaller wavelengths. The loss of intensity is described by modifying the scattering factor with a Debye–Waller factor

$$t(E,\theta)\exp\left(-|\boldsymbol{k}-\boldsymbol{k}'|^2(\Delta\boldsymbol{R})\right) = t(E,\theta)\exp(-2M) \tag{6.6}$$

where $|\boldsymbol{k}-\boldsymbol{k}'|$ is the momentum transfer for a particular diffraction beam and $\Delta\boldsymbol{R}$ represents the atomic displacements from their mean positions. For isotropic vibrations equation (6.6) becomes

$$\exp(-2M) = \exp\left[-\frac{3|\boldsymbol{k}-\boldsymbol{k}'|^2 T}{m_{\text{atom}} k_{\text{B}} \Theta^2}\right] \tag{6.7}$$

where T is the temperature, m_{atom} is the mass of the atom, k_{B} is the Boltzmann constant and Θ is the Debye temperature, which reflects the rigidity of the crystal lattice. It is clear that as the electron energy is increased the Debye–Waller factor decreases because the momentum transfer of a beam becomes larger and it is also evident that an increase in temperature will have the same effect. These effects limit quantitative LEED to energies $\leq 500\,\text{eV}$ since for higher energies the diffraction beam intensities become comparable to the level of the thermal diffuse background. Other factors will also limit the maximum usable LEED energies, such as the small separation of spots for low–order diffraction beams.

6.2.2 Intralayer Scattering

The scattering by a single atom has been described in the previous section in terms of the t-matrix (equation 6.5). The next stage of a LEED calculation involves calculating the multiple scattering from many atoms arranged in a single layer. The approach here will be to consider scattering between an increasing number of atoms and then to determine a scattering matrix for an entire layer, which could comprise several subplanes. These subplanes are normally used for adsorbates and complex structures with unit cells containing several atomic species. The detailed formalisms for the scattering and propagation of electrons within an atomic layer will not be dealt with here, but the interested reader is referred to the more detailed texts referenced at the end of the chapter. Here we consider only the basic principles of solving the multiple scattering problem.

First we consider the multiple scattering between two atoms in terms of the t-matrices of each atom and a set of Green functions that describe how spherical waves leaving one atom decompose into spherical waves arriving at another atom. A description of the use of Green functions in multiple scattering theory is beyond the scope of this book, and we will refer to them simply as propagator functions, P. Fig. 6.4(a) shows schematically the multiple scattering between two atoms in terms of the propagator functions. Since a spherical potential cannot affect angular

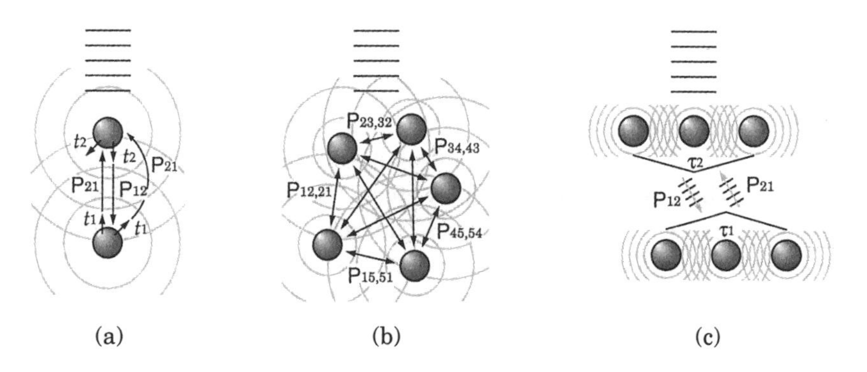

Fig. 6.4 An incident plane wave is decomposed into spherical waves and is scattered from (a) two atoms and (b) many atoms. (c) The intralayer scattering is simplified significantly by using the symmetry of the crystal and the interlayer scattering is usually represented by plane waves.

momentum, the amplitude of a spherical wave arriving at atom two after scattering from atom one is given by the product of the t-matrix for atom one and the propagator function, P_{21}. (Note that the subscript shows that the propagation is *to* atom 2 *from* atom 1.) This process can be extended for more scattering events such that the amplitude of a spherical wave after several scattering events is given by $t_2 \mathsf{P}_{21} t_1 \mathsf{P}_{12} t_2 \mathsf{P}_{21}$. Now we can derive expressions which give the amplitude of the spherical waves arriving at atom one and two after repeatedly scattering from one another

$$\begin{aligned} A_1 &= t_1 + t_1P_{12}t_2 + t_1P_{12}t_2P_{21}t_1 + t_1P_{12}t_2P_{21}t_1P_{12}t_2 + \ldots \\ A_2 &= t_2 + t_2P_{21}t_1 + t_2P_{21}t_1P_{12}t_2 + t_2P_{21}t_1P_{12}t_2P_{21}t_1 + \ldots \end{aligned} \tag{6.8}$$

For a LEED calculation, the propagator functions now have to be modified to express the difference in amplitude and phase of the external source of LEED electrons arriving at the two sites. This is done by introducing an amplitude ratio function into P_{12} and P_{21}. Equations (6.8) can be reduced, following Beeby (1968), to

$$\begin{aligned} A_1 &= t_1 + t_1P_{12}A_2 \\ A_2 &= t_2 + t_2P_{21}A_1 \end{aligned} \tag{6.9}$$

Such a reduction implies that the amplitude of the scattered waves at atom one is the sum of one scattering event from atom one plus all the scattering processes ending on atom two followed by propagation to atom one. A similar argument also holds for the amplitude of all the scattering processes ending on atom two. Using matrix notation, equation (6.9) can be written as

$$\begin{bmatrix} A_1 \\ A_2 \end{bmatrix} = \begin{bmatrix} I & -t_1P_{12} \\ -t_2P_{21} & I \end{bmatrix}^{-1} \begin{bmatrix} t_1 \\ t_2 \end{bmatrix} \tag{6.10}$$

This is the basic result that will be used to calculate all the multiple scattering within a layer after suitable extension to include many atoms ordered in a periodic plane. For the cluster of atoms shown in Fig. 6.4(b), equation (6.10) is easily extended, but in adding the quantities A_1, A_2, A_3, ... ,A_N, we must also account for the phase differences. The phase differences are calculated assuming an incoming and outgoing plane wave from the layer, which may seem peculiar considering that spherical waves have been used until now. However, since the scattering between layers is treated using a plane wave basis it is a natural choice.

Now that we are at a stage where we can calculate the multiple scattering from N atoms we could, in principle, begin to calculate the required diffraction intensities for the I–V spectra. However, the number of atoms would require the inversion of a prohibitively large matrix. Therefore, the next step is to incorporate the symmetry of the surface into the layer. In a single layer, whose symmetry is defined by the Bravais lattice, there is only one atom per primitive unit cell, as shown in Fig. 6.4(c). The scattering from all atoms is identical and so all the scattering amplitudes will be the same ($A_1 = A_2 = A_3 \ldots = A_N$). In this case, a new propagator function is defined that represents the sum of all multiple scattering within the layer. This propagator function can be combined with the t-matrix of the atoms within the layer such that

$$\tau = t\left(1 - t\sum_{n} P_{jn}\right)^{-1} \tag{6.11}$$

where the sum is over all atoms other than the jth atom. The τ-matrix represents the scattering of spherical waves from a symmetric arrangement of atoms in a similar manner to the t-matrix representation of atomic scattering. Now, by replacing the t-matrix in equation (6.10) with the τ-matrix for a Bravais layer the scattering from two planes can be calculated in an analogous manner. The addition of more layers with identical unit cells gives the scattered amplitudes from N layers as

$$\begin{bmatrix} A_1 \\ A_2 \\ \cdot \\ \cdot \\ \cdot \\ A_N \end{bmatrix} = \begin{bmatrix} I & -\tau_1 P_{12} & \ldots & -\tau_1 P_{1N} \\ -\tau_2 P_{21} & I & \ldots & -\tau_2 P_{2N} \\ \cdot & \cdot & & \cdot \\ \cdot & \cdot & & \cdot \\ \cdot & \cdot & & \cdot \\ -\tau_N P_{N1} & -\tau_N P_{N2} & \ldots & I \end{bmatrix}^{-1} \begin{bmatrix} \tau_1 \\ \tau_2 \\ \cdot \\ \cdot \\ \cdot \\ \tau_N \end{bmatrix} \tag{6.12}$$

where the subscripts on the propagator functions indicate scattering paths between the Bravais planes (Fig. 6.4(c)). The I–V spectra could now be calculated using equation (6.12) with the number of planes (N) equivalent to a few electron mean free paths. However, for most surfaces the matrices involved would still be far too large and so a more efficient means of combining the scattering from each layer is required. To this end the spherical–wave representation is transformed into a plane–wave representation and a diffraction matrix for the entire layer is defined. We recall that the layer could consist of subplanes so that the diffraction matrix would represent the scattered amplitudes from all the subplanes. In fact, four matrices must be defined to represent the transmitted and reflected amplitudes of an incident plane wave.

6.2.3 Interlayer Scattering

The multiple scattering problem was solved in the previous section since equation (6.12) can be used to calculate the diffraction intensities for any number of atomic layers. However, there are far more efficient ways of treating the scattering between the layers than by using the matrix inversion method of equation (6.11). A number of formalisms have been developed to calculate the interlayer scattering, including the layer–doubling (LD) method and the renormalised forward scattering (RFS) perturbation scheme (Pendry 1974). These two methods are used widely in quantitative LEED calculations; the former is useful for structures with small interatomic distances or relatively large penetration depths, whereas the latter is fast and efficient for most structures.

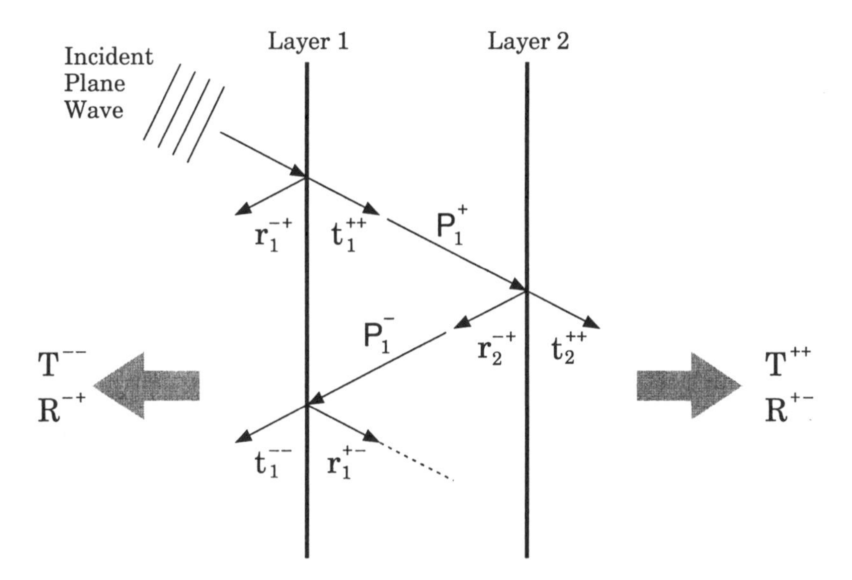

Fig. 6.5 The principal operations for calculating the interlayer scattering using the RFS perturbation scheme (Pendry 1974). This basic step is repeated for many layers and several cycles in and out of the crystal until the emerging intensity remains constant.

Consider the two layers shown in Fig. 6.5. These layers could consist of more than one subplane, but here we will assume that they comprise only one atomic plane. The incident plane wave is transmitted at the first layer and at the second layer is either transmitted or reflected with the amplitudes given by four matrices r^{-+}, t^{++}, r^{+-} and t^{--} (collectively denoted $m^{\pm\pm}$). These matrices are the result of solving the multiple scattering within the layer using equation (6.12) followed by changing the diffraction matrix from a spherical–wave basis to a plane–wave basis. By considering the sum over all paths four final matrices, R^{-+}, T^{++}, R^{+-} and T^{--} ($M^{\pm\pm}$) can be generated which are constructed from $m^{\pm\pm}$ and the plane wave propagators, P_1^- , P_1^+. By adding more layers and calculating the reflection and transmission matrices it is found that after ~ 10 layers the reflected amplitudes remain constant due to the limiting nature of the electron mean free path. In the LD scheme the fact that layers in the bulk are identical is exploited. Reflection and transmission matrices are calculated for two layers, the structure is doubled and $M^{\pm\pm}$ from the first step are used to calculate another set of reflection and transmission matrices for four layers. The next doubling will give $M^{\pm\pm}$ for eight layers in only double the time. The addition of a surface layer is straightforward once the bulk reflections have been calculated and stored so that surface relaxations and reconstructions can be determined with relative ease.

6.3 Reliability Factors

It is desirable to have an objective measure of the agreement between the experimental and calculated *I–V* spectra that is primarily sensitive to variations in the structural rather than the non-structural parameters. In LEED these are known as reliability factors, or r-factors, and provide a figure of merit for a proposed structure. Reliability factors quantify the agreement between *I–V* spectra, enabling the confidence in a structure determination to be maximised for a given experimental dataset. LEED *I–V* spectra are much more complicated than x-ray diffraction data; structural information is contained in the spectral peak profiles, intensities and energy positions. The energy positions of the peaks in an *I–V* spectrum are determined by the surface geometry. The intensities also depend on the surface geometry, but include effects due to inelastic losses and thermal vibrations and cannot be calculated as accurately as the peak positions. There are several r-factors used by LEED crystallographers that quantify the correspondence between I–V spectra based on the positions, asymmetries, widths and intensities of peaks. In the case of structure determinations of rare–earth metal surfaces, three r-factors have been widely used.

6.3.1 Zanazzi–Jona R-factor

One of the earliest r-factors for LEED crystallography was developed by Zanazzi and Jona (1977). Their r-factor, devised originally to establish the relationship between subjective visual evaluations and numerical evaluations of I–V spectra, takes into account the differences in both the gradients and the curvatures of the spectra. The Zanazzi–Jona r-factor for experimental (e) and theoretical (t) LEED I–V spectra from a single diffraction beam is given by

$$r_{ZJ} = \frac{N}{0.027}\int \frac{\left|cI_t'' - I_e''\right|\left|cI_t' - I_e'\right|}{\left|I_e'\right| + \left|I_e'\right|_{max}} dE \quad \text{where } N = \frac{1}{\int I_e dE} \tag{6.13}$$

The energy integrals are evaluated over the range of energies for which I–V spectra are available. The reduction factor of 0.027 reflects the mean value of r_{ZJ} for a random matching of any two unrelated beams and the factor c accounts for the scales in which the experimental (I_e) and theoretical (I_t) spectra are represented. To evaluate the reliability of a complete structure determination, involving comparisons of many diffraction beams, an energy–weighted average of all of the beams is taken to produce a final r-factor, R_{ZJ}. Values of R_{ZJ} less than ~ 0.2 indicate a high probability for a correct structure determination, whereas a value approaching 0.5 indicates a low confidence in the validity of the model under study.

6.3.2 Pendry R-factor

A very popular LEED R-factor was developed by Pendry (1980) in which equal weight is ascribed to all the peaks and troughs in the I–V spectra, eliminating any sensitivity to absolute intensities. The Pendry R-factor uses the logarithmic derivative ($L = I'/I$) of the I–V curves to define a

Y-function

$$\mathrm{Y} = \frac{L}{\left(1 + V_{\mathrm{oi}}^2 L^2\right)} \tag{6.14}$$

where V_{oi}, the imaginary part of the potential, is taken to be the average half–width of a single peak. The Pendry R-factor (a number between zero and two) is then given by

$$\mathrm{R_P} = \frac{\int (\mathrm{Y_e} - \mathrm{Y_t})^2 \mathrm{d}E}{\int \left(\mathrm{Y_e^2} + \mathrm{Y_t^2}\right) \mathrm{d}E} \tag{6.15}$$

6.3.3 Van Hove and Tong R-factor

In general, r-factors are designed to represent the correspondence between experimental and theoretical *I–V* curves using specific criteria. Different r-factors emphasise discrepancies in relative peak heights and energy positions to a different degree so that the r-factors themselves are subjective to some extent. In order to avoid this problem, Van Hove *et al* (1977) designed five r-factors, all of which are simultaneously minimised during a structure determination. The shortcomings of an individual r-factor were therefore reduced because each r-factor is sensitive to a range of differences between the experimental and theoretical spectra. The five r-factors are

$$\mathrm{r}_1 = \frac{\int |cI_\mathrm{t} - I_\mathrm{e}| \mathrm{d}E}{\int |I_\mathrm{e}| \mathrm{d}E}, \quad \mathrm{r}_2 = \frac{\int |cI_\mathrm{t} - I_\mathrm{e}|^2 \mathrm{d}E}{\int |I_\mathrm{e}|^2 \mathrm{d}E}$$

$$\mathrm{r}_3 = \text{fraction of energy slopes with opposite signs,} \tag{6.16}$$

$$\mathrm{r}_4 = \frac{\int |cI'_\mathrm{t} - I'_\mathrm{e}| \mathrm{d}E}{\int |I'_\mathrm{e}| \mathrm{d}E}, \quad \mathrm{r}_5 = \frac{\int |cI'_\mathrm{t} - I'_\mathrm{e}|^2 \mathrm{d}E}{\int |I'_\mathrm{e}|^2 \mathrm{d}E}$$

where c normalises the theoretical and experimental spectra to each other. The first two r-factors are sensitive to peak splitting, whereas the last three ensure that both peak slope and lineshape are matched. For the rare–earth structure determinations to be described in Chapter 7, the Van Hove and Tong R-factor (R_{VHT}) is given as a weighted average of the separate r-factors after minimisation.

6.3.4 Accuracy of a Structure Determination

One of the problems with R-factors is that it is difficult to make an estimation of the uncertainty in any structural parameter in a structure determination. The significance of any particular R-factor is related to the gradient of the 'valley' in parameter space. Pendry (1980) determined a significance factor based on the assessment of random fluctuations in the R-factor and also proposed a method of estimating the error in any parameter using a double reliability factor

$$\Delta R_P = \sqrt{\frac{8V_{oi}}{\Delta E}}\, R_P \tag{6.17}$$

where V_{oi} is the imaginary part of the potential and ΔE is the energy range of the experiment. Equation (6.17) is used to calculate the variation in the minimum R-factor which is then related to an error in the structural parameter determining the width of the R-factor valley ΔR_P. This procedure takes no account of multiple minima in the R-factor, which can affect the results significantly, and so must be used with caution.

However, no matter how accurately the multiple scattering of low–energy electrons is calculated, there is an inherent limit to the precision of a structural determination using LEED which is imposed by the diffraction process itself. The resolving power of the LEED method has been shown to depend on the energy range over which the data is taken and the instrumental resolution (Unertl and McKay 1984). In terms of the momentum transfer (Q), the resolving power is given by $Q_{max} - Q_{min}$ and the uncertainty in any particular measurement of Q so that

$$\frac{d}{\Delta d} = \frac{C}{2\pi n}\,\frac{Q_{max} - Q_{min}}{\Delta Q} \tag{6.18}$$

where d is the distance determined, C is a constant and n is the highest order of multiple scattering. The instrumental resolution is determined by the energy spread of the incident electron beam, the electron beam diameter, the angular dispersion of the electron beam and the aperture of the detector. These contributions cause uncertainties in the angle of incidence and in the scattered angle which result in a typical value of $1\,\mathrm{nm}^{-1}$ for ΔQ. The energy range of the data in a typical LEED experiment is limited at higher energies by the Debye–Waller factor (covered in Section 6.2.1). A lower limit is imposed by the geometry of the LEED optics so that the total energy range is usually ~30–300 eV ($Q_{max} - Q_{min} \approx 100\,\mathrm{nm}^{-1}$) for clean metal surfaces. If the above values are substituted into equation (6.18) we find a diffraction–limited resolution of $\Delta d/d \geq 0.5\%$. However, uncertainties in the scattering potentials and anharmonic surface vibrations can result in an uncertainty in structural parameters of $\Delta d/d \sim 2\%$.

6.4 Tensor LEED

The rare–earth metal surfaces which have been studied to date have not been complex enough to warrant the use of Tensor LEED (TLEED). However, TLEED may well have an important part to play in future studies of rare–earth metal surface reconstructions and investigations involving adsorbates. In this section, a brief outline of some of the advances made using TLEED will be discussed.

One of the difficulties with quantitative LEED is that the dynamical calculations discussed in Section 6.2 are performed repeatedly and compared to the experimental *I–V* spectra on a trial–and–error basis. The time taken to compute the matrix inversion that is required for the diffraction intensities scales as N^3, where N is the number of inequivalent atoms in the surface unit cell, and the time required for a complete trial–and–error search scales exponentially with the number of parameters varied. Advances in quantitative LEED have concentrated on reducing the time taken to calculate the diffraction intensities, which has been achieved by introducing new approximations in the multiple–scattering formalisms (Section 6.2.3). One such advance is TLEED, in which the dynamical calculations have been adapted to the needs of a directed–search strategy. In this respect, one fully dynamical calculation is performed for a given trial structure followed by rapid exploration of parameter space involving I–V spectra calculated using a perturbation scheme (Rous and Pendry 1989a,b). This requires an efficient means of calculating the I–V spectra at each step of a search and a procedure for continuing to the next step of the search based on a comparison between the theoretical and experimental spectra.

6.4.1 TLEED Theory

At the most basic level, the change in potential due to a distortion of the reference structure is calculated in terms of atomic displacements. A displacement of the jth atom by $\delta \boldsymbol{r}_j$ changes the surface potential by

$$\delta V_j \approx \delta \boldsymbol{r}_j \cdot \nabla V_j(\boldsymbol{r} - \boldsymbol{r}_j) \tag{6.19}$$

This distortion causes the amplitude of the scattered wave to be changed by

$$\delta \mathrm{A} \approx \sum_{j=1}^{N} \sum_{i=1}^{3} \mathsf{T}_{ij}\, \delta \boldsymbol{r}_{ij} \tag{6.20}$$

The tensor T_{ij} is summed over N displaced atoms and over the three Cartesian coordinates. If T_{ij} is determined for the reference structure using one fully dynamical calculation, then I–V spectra for many related trial structures can be determined by using equation (6.20), leading to a substantial increase in computational efficiency over conventional methods. The approach is accurate for atomic displacements comparable to about one tenth of the electron wavelength, which means a radius of convergence up to ~ 15 pm. A more sophisticated TLEED theory determines the change in the diffraction amplitudes using an angular momentum basis. In this approach the atomic t-matrix (Section 6.2.1) is expanded in terms of the atomic displacements

$$t_j^{'} = t_j + \delta t_j(\delta \boldsymbol{r}_j) \tag{6.21}$$

where δt_j is the change in the t-matrix of the jth atom produced by displacing it through $\delta \boldsymbol{r}_j$. The change in amplitude is then related to the atomic displacements through the tensor T (calculated only for the reference structure) and the set of δt. As the change in the surface potential due to the atomic displacements is determined more accurately, rapid calculations of *I–V* spectra for distortions of the reference structure can now be performed with a radius of convergence of ~ 40 pm.

6.4.2 Structure Determination

The aim of the structure determination is to find a global minimum on the R-factor hypersurface in parameter–space, as a function of the structural and non-structural parameters. There are several approaches to this depending on the complexity of the structure under investigation. In some cases, a coarse scan of the R-factor hypersurface is followed by a more detailed investigation around regions of local minima. In other cases, large jumps between randomly chosen regions of parameter space are made to find the global minimum. One of the shortcomings of all search algorithms is the fact that they sometimes get trapped in a local minimum. TLEED is therefore not a fully automated technique since manual input and assessment are still required at the end of a search.

One method that has been commonly used is the Marquardt search method (Marquardt 1963), but many other approaches have been implemented and some of these are discussed by Rous *et al* (1990) and Rous (1993).

Further Reading

Those interested in the detailed methodology of LEED and TLEED calculations are directed to the work of Pendry (1974) and Van Hove and Tong (1979). The optimisation of search strategies has been discussed by Van Hove *et al* (1993). The impact of TLEED upon surface crystallography has been reviewed more recently by Rous (1994).

CHAPTER 7

QUANTITATIVE LEED RESULTS

In general, rare–earth metal surface structures have been probed extensively using STM and surface x-ray diffraction, but here we will concentrate on the atomic structures obtained using quantitative LEED. In this respect, the (0001) and $(11\bar{2}0)$ surfaces of Gd and Tb have received considerable attention because of their interesting magnetic properties. For Gd, it has been argued that the (0001) surface has an enhanced Curie temperature compared to the bulk and that the surface may couple antiferromagnetically to the bulk (see Section 5.4.1). These observations stimulated several first–principles calculations which predicted that an expansion of the first interlayer spacing results in antiferromagnetic surface–to–bulk coupling (Wu RQ *et al* 1991a) and a contraction leads to ferromagnetic coupling (Eriksson *et al* 1995). The intimate relationship between magnetic phenomena and atomic coordination means that quantitative LEED provides vital structural information for testing theoretical calculations. The (0001) and $(11\bar{2}0)$ surfaces of Tb are interesting cases because of the relaxations and reconstructions that they exhibit. Since Tb thin films are ferromagnetic at low temperatures then similar correlations between magnetism and structure could enable further comparison with calculations involving partially occupied $4f$ minority states. The $(10\bar{1}0)$ surfaces exhibit extensive reconstructions, similar to the $(11\bar{2}0)$ surfaces, which shows that it is possible to create the basal plane structure from surfaces with different initial symmetries. The driving mechanism behind these reconstructions is currently not well understood; they may arise from purely surface free–energy considerations or be driven by small quantities of contamination at the surface or surface faceting reactions (Mischenko and Watson 1989a). In the case of the former, the observations would imply that the basal plane is the ground state surface for rare–earth metal surfaces, whereas the latter cases would imply the interesting possibility that rare–earth metal surfaces might be engineered to produce surfaces with varying symmetries.

7.1 (0001) Surfaces

To date, nearly all of the structural work on rare–earth metal surfaces has been performed on single crystals, but most spectroscopic work has been carried out on thin films grown on W(110) or Y(0001) substrates. Gd(0001) represents the only case where the two types of surfaces have been compared quantitatively using LEED *I–V*. However, the differences between single crystals and thin films have been discussed in a semi-quantitative manner using LEED *I–V* fingerprinting. In the following sections we will discuss the rare–earth metal surface structures determined using LEED *I–V* and compare the results with what is known about hcp surfaces in general and with first–principles calculations. To give an overview of the difficulty in preparing these surfaces, the cleaning procedure for each surface is described in some detail. As the first example, we will discuss the atomic structure of Sc(0001) since it is one of the few rare–earth metal surfaces for which the results of independent studies have been reported. The approach of Tougaard and Ignatiev (1982) for solving the Sc(0001) structure is also particularly useful for outlining the experimental LEED *I–V* procedure.

7.1.1 Sc(0001)

At the boundary between the 3*d* transition metals and the rare–earth metals, Sc is an interesting element. In many respects it behaves like a rare–earth metal, but its relative inertness makes the preparation of an atomically clean and well–ordered surface easier than for the lanthanide metals. The electronic structure of Sc is also intriguing since it provides a model rare–earth element without the complication of *f–d* hybridisation. Tougaard and Ignatiev (hereafter referred to as TI) first studied Sc(0001) due to an increasing interest in the hydrogen adsorption on this surface and because, at the time, relatively little was known about hcp (0001) surfaces. The cleaning procedure adopted by TI involved Kr ion sputtering at room temperature and at a series of elevated temperatures. After approximately 15 hours of ion sputtering at 900–1200 K, the initial surface contamination of C, S and Cl was removed, but a final residual contamination of oxygen at the level of 0.05 ML remained. The *I–V* spectra were recorded from a fluorescent screen using a spot photometer and the experimental spectra were compared to those calculated using the codes of Van Hove and Tong, which treated the intralayer scattering exactly and employed RFS for the interlayer multiple scattering (see Section 6.2.3). Since a number of hcp materials had already been reported to undergo hcp–to–fcc structural phase transitions, TI initially tested structural models which differed only in the layer registry. Fig. 7.1 shows the four possible surface terminations resulting in the hcp, fcc, fhcp and hfcc structures (the latter are defined in the figure caption).

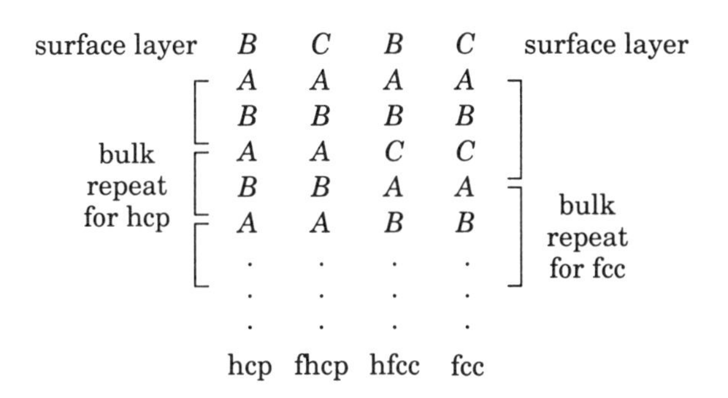

Fig. 7.1 Four possible terminations of a close–packed surface. The term 'fhcp' is used to describe a bulk hcp structure terminated by an fcc–like surface, and similarly 'hfcc' describes a bulk fcc structure terminated by an hcp–like surface.

For the ideal hcp structure it is evident that the (0001) surface can have either an *A* or *B* termination, which has already been discussed in Section 2.2.3.2. The equal probability of either termination increases the three–fold symmetry of the ideal surface to a six–fold symmetry for the real surface. Thus, all first–order diffraction beams have the same *I–V* profile, consider-ably reducing the number of independent *I–V* spectra available for the structural analysis.

Fig. 7.2 shows the experimental first–order LEED *I–V* spectrum obtained by TI together with the calculated spectra for the proposed structures. In this case a simple visual comparison clearly reveals that the bulk–terminated hcp structure gives the best agreement. To obtain a more accurate structural determination, TI also used off–normal incidence measurements, which breaks the experimental symmetry and so yields a much larger dataset. The recorded *I–V* spectra for the specular (00) beam,

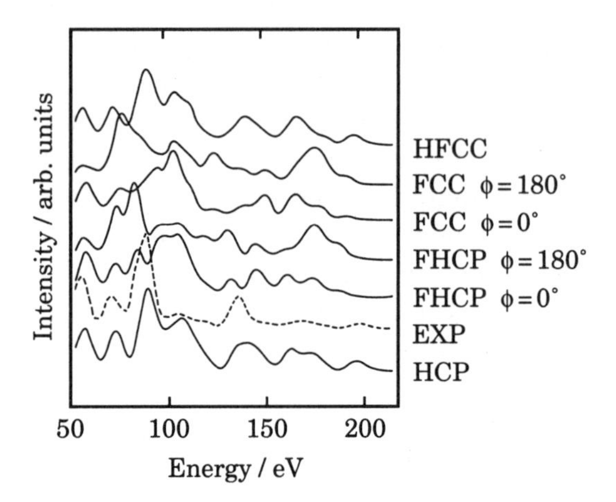

Fig. 7.2 LEED *I–V* curves for Sc(0001) calculated for the surface terminations shown in Fig. 7.1 compared with an experimental curve. Adapted from Tougaard and Ignatiev (1982).

and the first–order and second–order diffraction beams were then compared to the calculated spectra with the aid of the R_{ZJ}-factor. The results are summarised in Fig. 7.3, which shows the variation of the R_{ZJ}–factor as a function of the first interlayer spacing d_{12} and the inner potential V_i (see Section 6.2.1) – TI did not consider the relaxation of the second interlayer spacing, d_{23}.

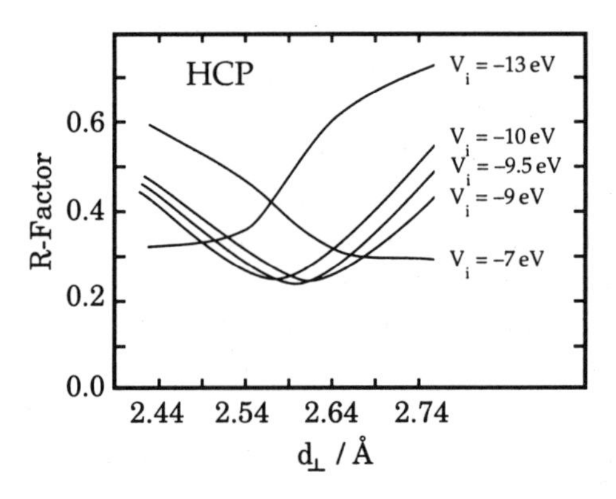

Fig. 7.3 Variation of R-factor as a function of first interlayer spacing for Sc(0001). Adapted from Tougaard and Ignatiev (1982).

The minimum R_{ZJ}–factor was found to be 0.23 for $d_{12} = 2.59 \pm 0.02$ Å with $V_i = -9.5$ eV, implying a 2 % contraction of the first interlayer spacing compared to the bulk value.

Following the prediction of an anomalous 5.1% expansion of the Sc(0001) surface by Chen (1992) (see Section 7.3), Dhesi *et al* (1995) also determined the Sc(0001) surface structure. The single–crystal surface was prepared by more than 30 cycles involving room–temperature Ar ion sputtering followed by annealing to 900 K, after which the surface was free of all contaminants except O (which was less than 0.05 ML). The experimental *I–V* spectra were recorded using a system based on a CCD camera interfaced to a computer frame–grabber (Barrett *et al* 1993) and the theoretical spectra were calculated using the CAVATN codes and the Tensor LEED approach (see Section 6.4). The initial structural search was conducted following the method of TI using first–order and second–order diffraction beams.

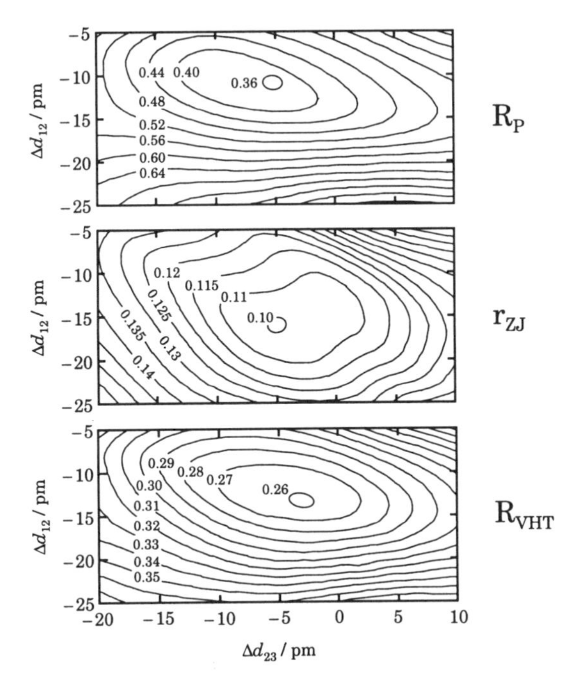

Fig. 7.4 Contour plots showing the variation of R-factors as functions of first and second interlayer spacings of Sc(0001). Adapted from Dhesi *et al* (1995).

The final search involved varying d_{12} and d_{23} for a hcp structure using the three R-factors described in Section 6.3. The variations of the R-factors with Δd_{12} and Δd_{23}, the changes in the values of d_{12} and d_{23}, respectively, are shown in Fig 7.4.

Table 7.1 Best–fit parameters for the surface structure of Sc(0001) and the corresponding minimum R-factors determined using the CAVATN and TLEED codes. Adapted from Dhesi *et al* (1995).

	Δd_{12} / Å	Δd_{23} / Å	V_0 / eV	R-factor	
	-0.11	-0.05	-7.0	0.36	R_P
CAVATN	-0.16	-0.05	-7.0	0.10	R_{ZJ}
	-0.13	-0.03	-7.0	0.26	R_{VHT}
	-0.10	-0.02	-5.0	0.34	R_P
TLEED	-0.24	+0.01	-9.0	0.10	R_{ZJ}
	-0.22	-0.01	-9.0	0.25	R_{VHT}

Table 7.1 shows the different parameters for the best–fit structures determined using the two calculational schemes. The final structure was given as $\Delta d_{12} = -0.16 \pm 0.03$ Å and $\Delta d_{23} = -0.03 \pm 0.03$ Å, representing a contraction of d_{12} by 6.1 % and a contraction of d_{23} by 1.1 %. The structure was verified using four off–normal–incidence *I–V* spectra (covering a total energy range of 950 eV) giving R-factor values of $R_{VHT} = 0.3$, $r_{ZJ} = 0.13$ and $R_P = 0.5$. For the first interlayer spacing the results are in qualitative agreement with those of TI, but in sharp contrast to the predictions of Chen (1992). Fig. 7.5 shows the experimental and calculated *I–V* spectra for the structure determined by Dhesi *et al* together with the *I–V* spectra calculated for the structure predicted by Chen. The correlation between the experimental *I–V* spectra and those corresponding to the structure of Chen is clearly poor, ruling out a 5 % expansion of the Sc(0001) surface layer.

7.1.2 Tb(0001)

The first lanthanide surface structure to be solved using quantitative LEED was that of Tb(0001) by Quinn *et al* (1991). The Tb ingot was refined using a lengthy SSE process, but considerable care had to be taken to monitor and reduce the level of contamination for the final experimental

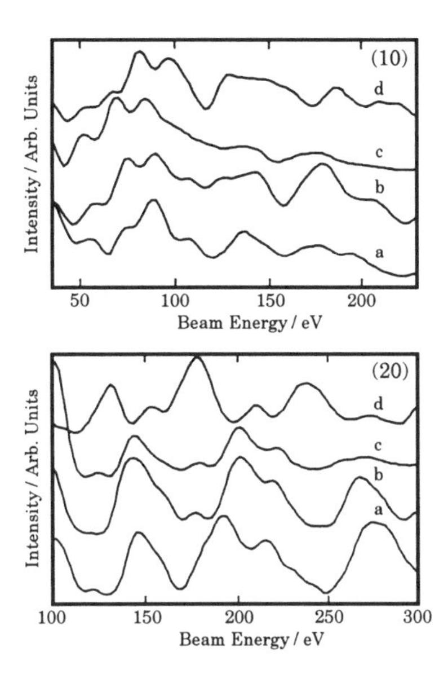

Fig. 7.5 LEED *I–V* spectra for Sc(0001): (a) Experimental spectra, (b) CAVATN calculation, (c) TLEED calculation, (d) CAVATN calculation for the structure proposed by Chen (1992). Adapted from Dhesi *et al* (1995).

surface. After 26 hours of Ar ion sputtering and annealing to 900 K the O level was reduced to below the AES sensitivity, but C and Cl remained. Ion sputtering with the sample held at 900 K followed by annealing at the same temperature gave an acceptable surface with the final levels of C and Cl being 6 % and 3 %, respectively. The data were collected using a television camera system that was controlled by a computer interface (Jona *et al* 1985). The calculations were performed using the Van Hove–Tong program invoking both RFS and layer–doubling. The inner potential was finally determined to be $-9 + 4.5i$ eV and the Debye temperature 188 K.

The structural search was initially performed for first–order and second–order diffracted beams over a large region of parameter space with the final refinement in a narrow zone around the best fit. The model assumed a bulk termination of the hcp structure from the start with the search then being conducted using variations in d_{12} and d_{23}. Fig. 7.6 shows the three R-factors as a function of the variation in Δd_{12} and Δd_{23}.

The final structure was determined to be $\Delta d_{12} = -0.11 \pm 0.03$ Å and $\Delta d_{23} = +0.04 \pm 0.03$ Å, giving a first interlayer contraction of 3.9% and a second interlayer expansion of 1.4%. The structure was then tested against seven off–normal–incidence *I–V* spectra (covering a total energy range of 1520 eV) which also gave quite good agreement with R-factor values of $R_{VHT} = 0.248$, $r_{ZJ} = 0.231$ and $R_P = 0.515$.

A first–interlayer contraction followed by a second–interlayer expansion fits in well with what is known about the structures of many hcp (0001) surfaces such as Ti, Zr and Re. One interesting exception is that of Be(0001), for which a 4.3% expansion of the first interlayer spacing has been found (Davis *et al* 1992) using a LEED *I–V* analysis. With the prediction of an expansion for the Gd(0001) surface (Wu RQ *et al* 1991a), mentioned at the beginning of this chapter, Quinn *et al* next turned their attentions to solving the Gd surface structure.

7.1.3 Gd(0001)

The Gd(0001) surface has received by far the most attention since the half-filled 4*f* gives the Gd atom a magnetic moment of 7 μ_B and the mediating conduction electrons help to establish ferromagnetic order very close to room temperature. However, recently the ferromagnetism of Gd has become a matter of debate with susceptibility measurements showing that needle–shaped Gd crystals exhibit an incommensurate order between 225 K and the Curie temperature of 292 K. The magnetic properties of Gd thin film surfaces have, however, attracted much more attention since the discovery of a surface–enhanced Curie temperature by Weller *et al* (1985a). Using SPLEED and SPUPS for surface–sensitive measurements and MOKE for bulk measurements, Weller *et al* found that the surface Curie temperature was enhanced by 22 K and that the surface coupled antiferromagnetically to the bulk. The antiferromagnetic coupling was later confirmed by Wu RQ *et al* (1991a) using first–principles calculations, which also predicted a 6.3% expansion of the first interlayer spacing compared to the bulk. In contrast, further studies showed that the Curie temperature was enhanced by 60–85 K, but that the surface coupled ferromagnetically to the bulk (Tang H *et al* 1993a). Thus, for some time it seemed that the Gd(0001) surface exhibited highly exotic and interesting magnetic properties which required much further study. After the work of Tang H *et al* (1993a), more first–principles calculations

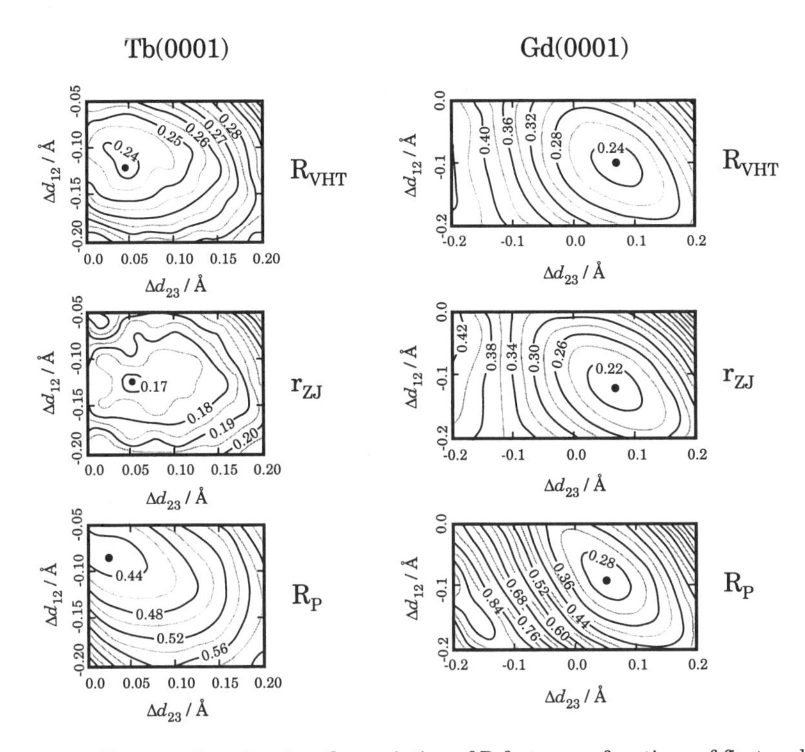

Fig. 7.6 Contour plots showing the variation of R-factors as functions of first and second interlayer spacings of Tb(0001) and Gd(0001). Adapted from Quinn *et al* (1991) and Quinn *et al* (1992), respectively.

supported a ferromagnetic surface–to–bulk coupling and a first–interlayer spacing contraction of 3.5–4.5 % (Eriksson *et al* 1995).

The motivation behind the first quantitative study of Gd(0001) was thus to establish its surface structure to aid first–principles calculations in determining its magnetic properties. This study was performed on a single–crystal surface, but an alternative and easier method to obtain clean Gd(0001) surfaces involves the growth of thin films on W(110) or Y(0001) substrates. The motivation behind the second quantitative study of Gd(0001) was therefore to test if thin films grown on W(110) have the same structure as single–crystal surfaces. The Gd surface therefore represents one of the few cases in rare–earth surface science where the combined use of an array of experimental and theoretical approaches has lead to a much more detailed understanding of the surface.

The first structural determination of Gd(0001) was performed by Quinn *et al* (1992) on a single–crystal surface prepared by 39 hours of Ar ion sputtering and a one–hour anneal. Though the sputtering reduced all contaminants to below the AES sensitivity level, the annealing resulted in C and Fe segregation to the surface. The cleanest surface was produced by 25 minutes of Ar ion sputtering, heating to 1350 K in one minute for a total of two minutes and then cooling rapidly to 650 K in 90 seconds. This procedure resulted in a surface with acceptably low levels of C, O, Cl and Fe. Quinn *et al* used the same analysis program as for the Tb(0001) structural determination with a final inner potential of $-9+4.5i$ eV and a Debye temperature of 152 K. All models tested were based on a bulk termination of the hcp structure with relaxations of d_{12} and d_{23}. Fig. 7.6 shows the dependence of the three R-factors on Δd_{12} and Δd_{23}. The final result was quoted as $\Delta d_{12}=-0.1\pm0.03$ Å and $\Delta d_{23}=+0.06\pm0.03$ Å, representing a 3.5 % contraction of d_{12} and a 2 % expansion of d_{23}. These values were further confirmed with off–normal–incidence spectra from seven

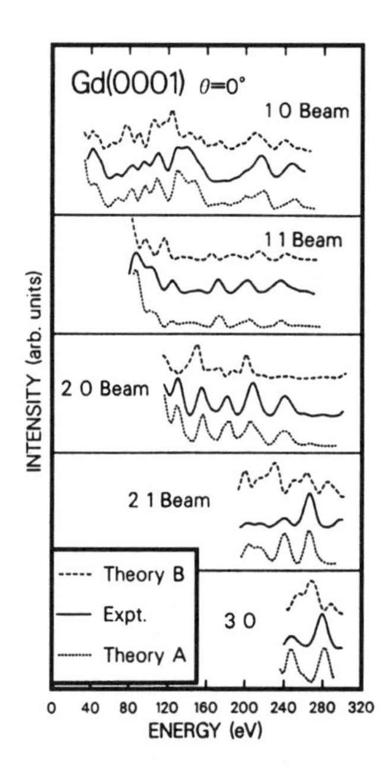

Fig. 7.7 LEED *I–V* spectra for normal incidence on Gd(0001): Solid curve is experimental; Theory A is the best–fit model; Theory B is calculated for an expansion of the first interlayer spacing of 6.3 %. Adapted from Quinn *et al* (1992).

diffraction beams (covering a total energy range of 1410 eV) which gave R-factors of $R_{VHT} = 0.22$, $r_{ZJ} = 0.2$ and $R_P = 0.25$. Fig. 7.7 shows the normal–incidence experimental spectra, together with the *I–V* spectra for the best fit structure and those for the structure predicted by Wu RQ *et al*. Clearly a visual comparison of the spectra rules out a 6.3 % expansion of the surface. The predictions of all other first–principles calculations agree quite well with the results of Quinn *et al* (1992) and are discussed further in Section 7.3.

Since most magnetic studies of rare–earth metal surfaces are performed on epitaxially grown thin films, it is important to establish their structure using precise atomic probes. This also opens up the possibility of comparing the structural properties of thin film surfaces grown under different conditions with those of single–crystal surfaces. Giergiel *et al* (1995, 1996) have grown 40 nm Gd films on W(110) at room temperature, followed by a short annealing to 875 K, and also at elevated temperatures of 825 K. The *I–V* spectra were analysed with the Van Hove and Tong multiple–scattering code with a final inner potential of $-6 + 4i$ eV and a Debye temperature of 176 eV, which are slightly different to the parameters reported by Quinn *et al* (1991).

Table 7.2 Optimal structure and R-factor values for Gd films grown at low (300 K) or high (825 K) temperature. Adapted from Giergiel *et al* (1995).

	Low temperature			High temperature		
	R	Δd_{12} / Å	Δd_{23} / Å	R	Δd_{12} / Å	Δd_{23} / Å
R_P	0.20	-0.08	+0.03	0.20	-0.11	+0.02
R_{ZJ}	0.34	-0.07	+0.03	0.37	-0.07	+0.03
R_{VHT}	0.23	-0.07	+0.03	0.27	-0.07	+0.03

Table 7.2 gives the final structure determined using all diffraction beams up to second order and shows that there is essentially no difference between the two growth modes adopted by Giergiel *et al*. By averaging all the structures over different R-factors Giergiel *et al* finally quote a structure with $\Delta d_{12} = -0.07 \pm 0.01$ Å and $\Delta d_{23} = +0.03$ Å, representing a contraction of d_{12} by 2.4 % and an expansion of d_{23} by 1 %. This compares very well with the results for single–crystal surfaces, indicating that Gd thin films grown on W(110) have a very similar structure to single–crystal surfaces.

7.1.4 Fingerprinting

A full structural search, as in the cases of Tb(0001) and Gd(0001) described previously, provides one means of comparing surfaces prepared using different methods, but *I–V* spectra can also be used to make a semi-quantitative comparison. The *I–V* spectra, while needing careful analysis to gain full structural information, can be used as 'fingerprints' for a particular surface structure. In this manner Li H *et al* (1992) compared the *I–V* spectra for single–crystal and thin–film surfaces of Tb(0001) and also recorded *I–V* spectra for Gd, Dy, Ho and Er. Fig. 7.8 shows the normal–incidence *I–V* spectra for the Tb(0001) surfaces and it is reasonable to conclude from such data that the single–crystal and thin–film

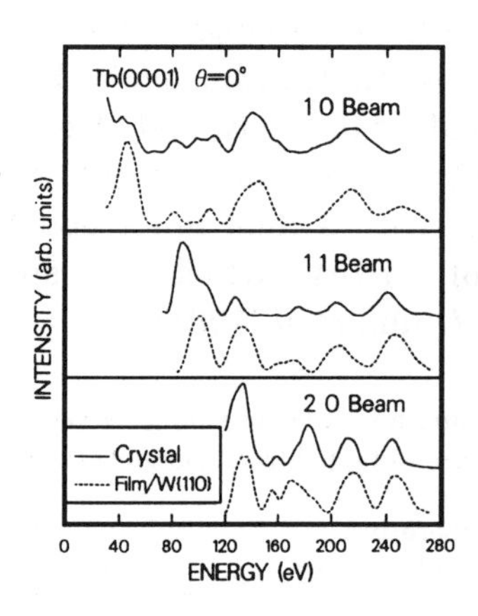

Fig. 7.8 Comparison of normal incidence LEED *I–V* spectra from (solid) bulk Tb(0001) and (dashed) a Tb(0001) thin film. Adapted from Li H *et al* (1992).

surfaces are, at least qualitatively, similar in structure. However, based on normal–incidence photoemission from the same surfaces, Li H *et al* further concluded that the long–range order in the thin films is less well developed than for bulk single crystals. This was based on a reduced surface–state contribution at the Fermi level and broadened 4*f* peaks. Other workers have reported much stronger surface–state features for rare–earth metal thin films and have resolved small surface core–level shifts in the 4*f* features of Tb (see Section 5.3.2). Thus it would still seem that rare–earth metal thin films offer a convenient alternative for preparing high–quality single–crystal surfaces.

7.2 (11$\bar{2}$0) Surfaces

The (11$\bar{2}$0) surfaces of hcp structures are intriguing because, unlike (0001) surfaces, the surface unit cell contains two inequivalent atoms (see Section 2.2.3.1) which can relax in different ways and still retain the symmetry of the surface. Prior to the study of Tb (11$\bar{2}$0) by Li YS *et al* (1992) the only hcp (11$\bar{2}$0) surface to have been studied was Co (11$\bar{2}$0), which showed no lateral displacement of the inequivalent sites.

The (11$\bar{2}$0) surface also provides interesting possibilities in spectroscopic probes of the band structure. The study of electronic structure using ARUPS involves transitions from occupied to unoccupied states which inherently destroys the momentum conservation perpendicular to the surface (Himpsel 1982). One way to regain the lost information is to acquire ARUPS data from different crystallographically orientated surfaces and then use a triangulation technique to resolve all the momentum components of the initial state. Unfortunately, this approach failed to yield the desired results for the rare–earth metals because of a significant reconstruction of the surface (see Section 5.2.1.1). However, other workers have reported a bulk–like termination of the (11$\bar{2}$0) surface and it is this work in which we are primarily interested due to its quantitative nature.

The differences in the reported LEED patterns from rare–earth (11$\bar{2}$0) surfaces may well lie in the different approaches to obtaining an acceptably clean surface. Whereas Barrett (1992a) used repeated cycles of Ar ion bombardment followed by 30 minutes of annealing at 900 K, Li YS *et al* (1992) and Quinn *et al* (1993) adopted a dramatically different approach. After four cycles of 20 hours of Ar ion sputtering at 900 K followed by a one–hour anneal at 900 K, the Cl was eliminated, but C, O and Fe were still detectable using AES. To obtain an surface free of Fe contamination and reduce the levels of C and O to acceptable levels, the surface was sputtered for 30 minutes, annealed at 1400 K for 10 minutes, cooled to 850 K in less than two minutes, and subsequently annealed for three hours at 700 K. For Tb (11$\bar{2}$0), this approach gave the sharp (1 × 1) LEED pattern shown in Fig. 7.9a with spots being visible up to electron energies of 300 eV. The schematic of the LEED pattern (Fig. 7.9b) shows that the unreconstructed surface is characterised by missing diffraction beams due to a glide line (Fig. 7.9c). In a kinematic approach to diffraction from an ideal (11$\bar{2}$0) surface more extinctions would be

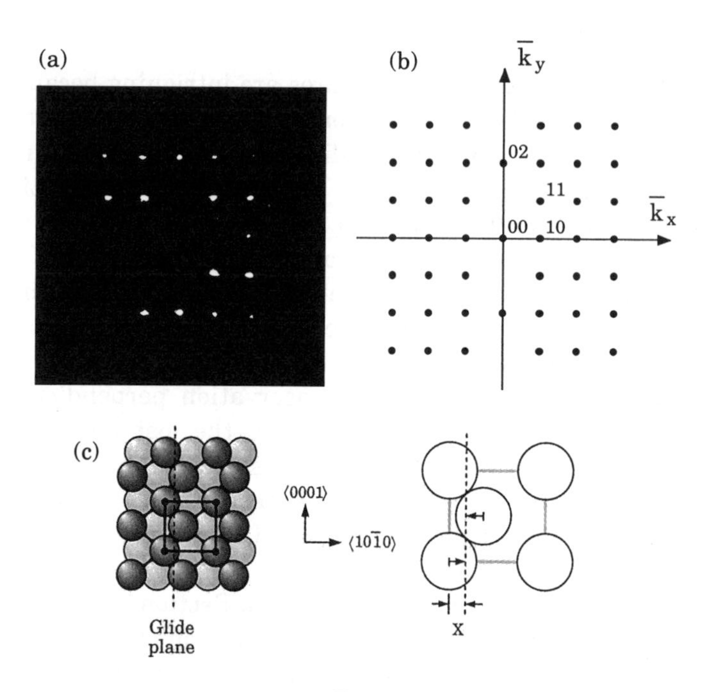

Fig. 7.9 (a) LEED pattern from Tb (11$\bar{2}$0). (b) Schematic LEED pattern showing extinction of the (01) spot. (c) Ideal (11$\bar{2}$0) surface and definition of the parameter labelled X. Adapted from Li YS *et al* (1992).

expected, but these diffraction spots are reported to have a finite intensity in the experiment. One reason for this is that dynamical diffraction can give a non-zero intensity even for an ideal bulk–terminated surface. A second reason is that if the position of the inequivalent atoms in the surface unit cell move uniaxially about the glide line, then the extinction conditions is violated for the axis of displacement.

The LEED intensities were calculated using the CHANGE program of Jepsen (1980) with a final inner potential of $-7 + 4i$ eV and a Debye temperature of 170 K. The experimental I–V spectra were recorded for 11 diffracted beams at normal incidence covering a total energy range of 1440 eV. The structural search was conducted by changing d_{12} and the relative position of the second atom of the two–atom basis, giving the search parameter X (defined in Fig. 7.9c). For the purposes of the calculation the surface was defined as two interlocking rectangular unit cells which were moved relative to each other. The search was quantified using three R-factors whose variation with d_{12} and X is shown in Fig. 7.10. The final structure was $\Delta d_{12} = -0.06 \pm 0.03$ Å and $\Delta X = +0.21 \pm 0.05$ Å, representing a 3.3 % contraction of d_{12} and a 20 % increase in X. Li YS *et al* (1992) thus observed the first lateral surface relaxation of a hcp surface where the registry of two subplanes in the surface layers is changed, but the symmetry maintained.

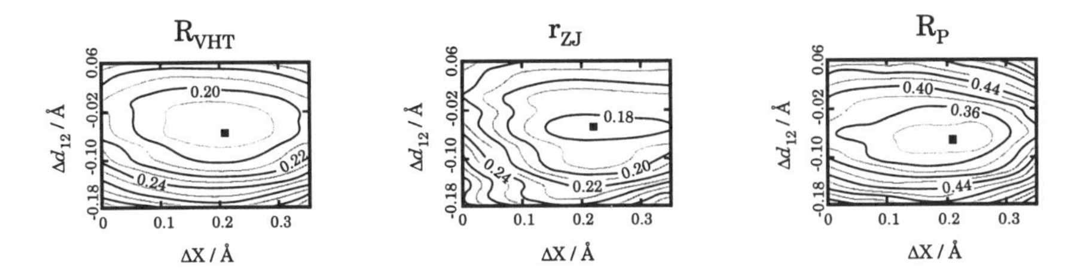

Fig. 7.10 Contour plots showing the variation of R-factors as functions of first and second interlayer spacings of Tb (11$\bar{2}$0). Adapted from Li YS *et al* (1992).

Only one further study of a rare–earth metal (11$\bar{2}$0) surface has been reported. Using the same procedure as that described above for Tb, Quinn *et al* (1993) studied Gd (11$\bar{2}$0) and found $\Delta d_{12} = -0.05 \pm 0.03$ Å and $\Delta d_{23} = -0.02 \pm 0.03$ Å, representing contractions of 2.7 % and 1 %, respectively. However, for Gd (11$\bar{2}$0) the increase in X was less conclusive than for Tb, being only $+0.1 \pm 0.1$ Å, representing an increase of $\Delta X \sim 10$ %.

7.3 Calculated Surface Structures

The interface energy calculation, mentioned in Section 5.2.2 in the context of interpretation of RHEED data, is a example of an energy–minimisation approach to surface structure determination. An overlayer and substrate layer, both of known structure, are moved with respect to each other and the interface energy is calculated from the potentials between the constituent atoms. The minimum value of the interface energy gives the most probable registry between the overlayer and the substrate. Clearly, although this approach can provide useful information on adsorption sites in overlayer structures, it does not enable the calculation of the structure of the surface itself. Of more interest here are the types of calculations that provide more direct information on the surface structure, through minimisation of the calculated total energy of a system by changing various structural parameters, such as the extent to which a surface is relaxed from an ideal bulk–terminated structure.

The surfaces of hcp metals have, in general, received far less attention in terms of first–principles structural calculations than, for instance, fcc and bcc metal surfaces. One of the most extensive studies investigated 13 hcp metals using local volume potentials (LVP, a method similar to the embedded–atom method). On the basis of these calculations, Chen (1992) proposed that the expansions and contractions of the first interlayer spacings of hcp metals are element–specific and not intrinsic properties of particular surfaces. The same calculations for the related (111) surfaces of fcc metals showed that a contraction of the first interlayer spacing is a common feature of these surfaces. In the LVP formalism, a pair–wise interaction attracts the surface atoms towards the bulk whereas another term representing many–body interactions always repels the surface away from the bulk. Chen argued that in the case of Sc, Dy and Er the many–body repulsions are stronger than the pair–wise attractions, which results in the corresponding expansion of the first interlayer spacing. However, the expansion of the surface is not confirmed for Sc, indicating that the magnitude of the two competing effects may have to be reconsidered or that other terms may have to be included in the expansion of the surface free energy.

The calculations performed for the Gd surface have been mentioned already in Section 7.1.3 with regards to the experimental results. Wu RQ *et al* (1991a) performed the first extensive theoretical analysis of the Gd surface using full–potential linearised augmented–plane–wave energy band calculations to study the electronic, magnetic and structural properties of the Gd(0001) surface. These calculations predict a surface state of d character near the Γ point of the Brillouin zone which has been observed in many photoemission experiments. Fig. 7.11 shows the total energy calculated for the Gd(0001) surface as a function of the first interlayer spacing and the surface–to–bulk coupling. Based on these results Wu RQ *et al* concluded a first interlayer expansion of 6.3% and an antiferromagnetic coupling of the surface to the bulk. Wu RQ *et al* argued that the expansion of the surface was a result of competition between surface hybridisation with underlying layers and the exchange–correlation energy bases of the LSDA. The former tends to lower the first interlayer spacing to saturate the dangling bonds at the surface, but the latter

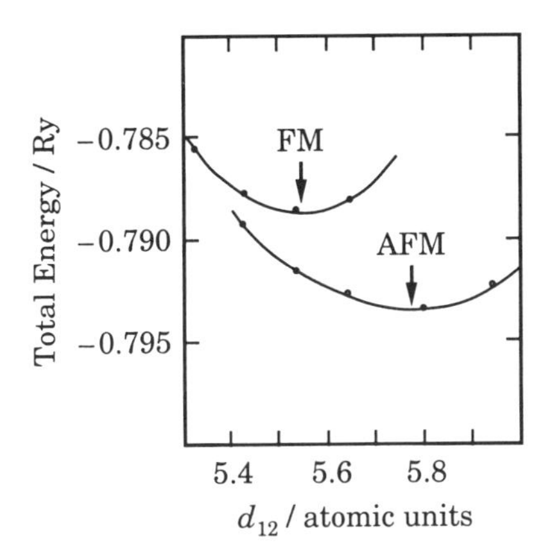

Fig. 7.11 Total energy of the Gd(0001) surface as a function of the first interlayer spacing and surface–to–bulk coupling (FM = ferromagnetic, AFM = antiferromagnetic). Adapted from Wu RQ *et al* (1991a).

favours larger spacings. However, as was stated in Section 7.1.3, experimental results do not confirm these predictions of Wu RQ *et al*. Jenkins and Temmerman (1999) applied the LSDA to calculations of the coupling in bulk Gd and found that it is very sensitive to the interlayer spacing. As this spacing changes at the surface due to surface relaxation, these calculations predict that different surface–to–bulk coupling will result from small changes in surface relaxation, with ferromagnetic coupling being favoured for surface interlayer spacings of less than the bulk value of ~ 0.29 nm and antiferromagnetic coupling for values greater than this.

Bylander and Kleinmann (1994) pointed out that the large expansion of the surface layer calculated by Wu RQ *et al* may have resulted from allowing only the surface layer the freedom to move away from the bulk. They then argued that bulk Gd calculations would have given a lattice constant 2–3 % larger than the experimental value, and suggested that in the thin film calculations the lattice constants may have been reduced to the experimental values, leaving only the surface interlayer spacing to relieve the pressure. They also argued that the antiferromagnetic coupling was simply a failure of the LSDA, since Heinemann and Temmerman (1994) had already reported that without gradient corrections the LSDA calculations predict even bulk Gd to be antiferromagnetic! In the approach of Bylander and Kleinmann (1994) the occupied 4*f* electrons were treated as localised states whereas the unoccupied $4f$ states were assumed to have band character. By using a single variable fitting parameter to adjust the calculated Gd moment with the experimental value, they predicted a ferromagnetic surface–to–bulk coupling and also reproduced accurately the experimentally observed first interlayer contraction and second interlayer expansion reported by Quinn *et al* (1992). In another approach, Eriksson *et al* (1995) treated both the unoccupied and occupied $4f$ states as core levels, excluding hybridisation with the valence electrons in a similar manner to that of Wu RQ *et al* (1991a). With this assumption, Eriksson *et al* showed that, even without gradient corrections to the LSDA or a variable fitting parameter, the surface of Gd(0001) is calculated to be ferromagnetically coupled to the bulk and that the first interlayer spacing is contracted by 4.4%.

Thus, the surface structure and magnetism of Gd appears to be accessible using a first–principles approach within the LSDA approximation and fitting parameters are unnecessary. The reason for the discrepancy between the results of Wu RQ *et al* and Eriksson *et al* are unclear, but the latter authors point to the different treatments of the $5p$ states as a possible source. Tables 7.3–7.5 at the end of this Chapter summarise the current experimental and theoretical understanding of rare–earth metal surface structures for the (0001), $(11\bar{2}0)$ and $(10\bar{1}0)$ surfaces, respectively. At present, only quantitative LEED has been used to determine the atomic dimensions of the surfaces, and so comparison with other techniques such as ion and He scattering is highly desirable.

7.4 Outlook

The overview of rare–earth metal surface structure determinations presented in this chapter shows that only in the case of Gd(0001) have theoretical calculations and experimental results been combined to give a deeper understanding of the surface structure. Given the importance of the LSDA in electronic structure calculations, rare–earth metal surfaces present ideal cases where this theoretical framework can be tested and improved, as has been clearly demonstrated in the case of Gd. The conflicting reports for the $(11\bar{2}0)$ surfaces remain to be resolved, but the reconstructions and relaxations of this surface should stimulate much more experimental and theoretical activity. In this respect, quantitative LEED measurements from the extensively reconstructed $(11\bar{2}0)$ surface should provide another useful test for the TLEED approach to solving surface structures. The $(10\bar{1}0)$ surfaces for some elements have been calculated (Table 7.5), but experimental results are lacking for these surfaces. The major obstacle to an improved understanding of more rare–earth metal surfaces seems to be the difficulty in preparing atomically clean and well–ordered surfaces. With current advances in crystal purification techniques and uhv surface preparation methods, more structural determinations should follow. It can be seen from the results and predictions in Tables 7.3–7.5 that calculating the structure of rare–earth metal surfaces is still largely unexplored territory and that these surfaces present an excellent testing ground for future theoretical models.

Table 7.3 Summary of surface structure determinations for (0001) surfaces.

		Experiment			Theory		
	Bulk d / Å	Δd_{12} / Å	Δd_{23} / Å	Ref	Δd_{12} / Å	Δd_{23} / Å	Ref
Sc	2.63	-0.05 ± 0.02 (-2%)		Tougaard	+0.132 (+5.1%)	+0.02 (+0.8%)	Chen
		-0.16 ± 0.03 (-6.1%)	-0.03 ± 0.03 (-1.1%)	Dhesi			
Gd	2.89	-0.10 ± 0.03 (-3.5%)	+0.06 ± 0.03 (+2.0%)	Quinn 92	+0.182 (+6.3%)	+0.064 (+2.2%)	Wu RQ
		-0.07 ± 0.01 (-2.4%)	+0.03 (+1.0%)	Giergiel	-0.066 (-2.3%)	-0.027 (-0.9%)	Chen
					-0.098 (-3.4%)	+0.048 (+1.7%)	Bylander
					-0.127 (-4.4%)	0	Eriksson
Tb	2.85	-0.11 ± 0.03 (-3.9%)	+0. 04 ± 0.03 (+1.4%)	Quinn 91	–	–	
Dy	2.83	–	–		+0.048 (+1.5%)	-0.245 (-0.8%)	Chen
Er	2.79	–	–		+0.054 (+1.9%)	+0.02 (+0.1%)	Chen

Table 7.4 Summary of surface structure determinations for $(11\bar{2}0)$ surfaces.

	Bulk d / Å	X / Å	Δd_{12} / Å	Δd_{23} / Å	ΔX / Å	Ref
Gd	1.82	1.05	-0.05 ± 0.03 (-2.7%)	-0.02 ± 0.03 (-1%)	+0.1 ± 0.1 (+10%)	Quinn 93
Tb	1.80	1.04	-0.06 ± 0.03 (-3.3%)	–	+0.21 ± 0.05 (+2%)	Li YS

Table 7.5 Summary of surface structure predictions for $(10\bar{1}0)$ surfaces.

	Bulk d / Å	Δd_{12} / Å	Δd_{23} / Å	Ref
Sc	0.95	+0.089 (+9.3%)	+0.073 (+3.8%)	Chen
Gd	1.05	-0.089 (-8.5%)	-0.003 (-0.2%)	Chen
Dy	1.04	+0.014 (+1.3%)	+0.068 (+3.3%)	Chen
Er	1.03	-0.006 (-0.6%)	-0.044 (-2.1%)	Chen

CHAPTER 8

SUMMARY
PAST, PRESENT AND FUTURE

The Past

The surface reconstructions observed on bulk single–crystal samples of the rare–earth metals are still the most unusual reconstructions of any surface yet studied. The depth to which the surfaces are totally reconstructed is greater than that which can be probed with many surface–sensitive techniques, preventing progress in the determination of the subsurface structures. The concomitant change of rotational symmetry of the surface, from two–fold to three–fold, remains a challenge to theoretical explanations of the reconstructions.

The enigmatic surface–order dependent state observed in photoemission spectra has also evaded a satisfactory explanation throughout the past decade. There are signs of some progress being made, but whether it is characteristic of a clean surface or a side–effect of an interaction between surface contaminants, a mechanism explaining its origin is still elusive.

Studies of the epitaxial growth of thin films of rare–earth metals on refractory metal substrates have come to a consensus regarding the growth modes. With the substrate held at room temperature the films exhibit FM growth, at least for a few monolayers, whereas elevated temperatures promote SK growth. Annealing flat films after growth, to improve the crystallographic quality by reducing the defect concentration, produces islanding if the temperature is taken above a critical value that increases with increasing film thickness. In this case, the resultant film morphology is similar to those films grown in the SK growth mode.

Calculations of the surface relaxation and surface–to–bulk magnetic coupling for Gd have managed to agree with the experimental data that existed at the time, regardless of whether that data indicated antiferromagnetic or ferromagnetic coupling. Future experiments will no doubt reveal yet more detail that the theorists will succeed in explaining subsequently in terms of the latest theoretical ideas.

The Present

Interest in the valence instability of some of the rare–earth metals and compounds, which is an intrinsically bulk property that is influenced by the presence of the surface, started back in the 1960s. Even after four decades, it is still a subject of debate amongst experimentalists and theorists.

Surface magnetism continues to be an active area of research, with SEMO and surface–to–bulk coupling looking set to remain hot topics for a few years to come. The increase of the extent of SEMO has been attributed to improvements in the quality of the epitaxial thin films grown, and so it remains to be seen if the values reported recently can be improved upon. Studies of rare–earth metals other than Gd and Tb are being made in the hope of elucidating some of the systematics of the magnetic properties of the lanthanides.

The Future

What does the future hold for rare–earth surface science? We are still a long way from understanding all of the subtleties of the interconnections between the geometric, the electronic and the magnetic structures of rare–earth metal surfaces. Whether we will see, for instance, SEMO applied to magnetic recording technology before a full understanding has developed is an interesting proposition.

The hardware and software required to do a full first–principles calculation of the equilibrium structure of the surface reconstructions, within a sensible timescale, are not readily available. Not quite. The processing power of computers is increasing exponentially, so we can hope for this capability early in the next decade.

The epitaxial growth of metal–on–metal thin films has concentrated almost entirely on systems comprising close–packed atomic layers. There is still much work that can be done on more open structures, including strained layers and growth on substrates that promote hcp non-basal–plane layered growth. This may see the largest expansion in the next few years.

The use of STM in surface characterisation should help to improve the reproducibility of thin–film preparation and so enable better correlation of the results from different research groups. Maybe we will see STM become as ubiquitous as LEED on surface science uhv chambers in the laboratory. In the field of surface magnetism, the use of STS will surely become more popular as a result.

If the current trend continues, away from 'artificial' surfaces in uhv chambers and towards 'real' surfaces in more hostile environments, then dealing with reactive rare–earth metal surfaces could prove to be problematic. This does not preclude the use of surface–sensitive techniques that do not require a solid/vacuum interface, such as optical probe spectroscopies. Such techniques can be developed to study nanoscale phenomena in rare–earth metal structures protected beneath inert buffer layers, thus extending their application to environments that are, as yet, unexploited.

APPENDIX

TABLES OF RARE–EARTH METAL SURFACE SCIENCE STUDIES

Tables A.1 – A.4 list the experimental and theoretical studies that have been carried to date (January 1999) on rare–earth metal surfaces. For the purposes of clarity the studies have been separated into two categories.

The first category covers the surfaces of bulk single crystals and epitaxial thin films of thickness greater than ~ 1 nm, or ~ 4 ML. For epitaxially grown films thicker than ~ 1 nm it is often assumed, either implicitly or explicitly, that the substrate has little or no influence on the properties of the rare–earth surface being studied. The studies are tabulated by technique if the specific focus of the study is the surface structure (Table A.1), but studies of the electronic and magnetic structures are labelled under those generic headings, regardless of whether or not information on the surface geometric structure is inferred (Table A.2).

The second category covers ultra-thin films of thickness less than ~ 1 nm. For such films it is understood that the substrate material and its crystalline structure will have significant influences on the films grown. Studies of ultra-thin films are tabulated by the substrate material, separated into refractory metals (Table A.3) and transition metals and other materials (Table A.4).

Some studies include films of various thicknesses that span both regimes, and so these are included in both categories.

Table A.1 Studies of single–crystal rare–earth metal surfaces (geometric structure)

	LEED	RHEED	STM	SXRD	XPD	Theory
Sc	B5 D1 T4	–	D1	–	–	C1
Y	B1 B3 B4 B6	T1 T2	R1 T2	–	H1	–
La	–	–	–	–	–	–
Ce	–	–	–	–	–	–
Pr	B6	–	–	–	–	–
Nd	–	–	–	–	–	–
Sm	S2	–	–	N2	–	–
Eu	–	–	–	–	–	–
Gd	G1 L1 M1 Q2 Q3 W1 W2 W3 W4	F1	K1 T3 W2 W4	M2	–	B7 C1 E1 W5
Tb	L1 L2 Q1 S1	K2	K1 K2	–	–	–
Dy	L1 S1	Y1	–	–	–	C1
Ho	B2 B6 L1	–	–	–	–	–
Er	B2 B6 L1	–	–	–	–	C1
Tm	N1	–	–	–	–	–
Yb	–	–	–	–	–	–
Lu	–	–	R1	–	–	–
Ln	B4 B6	–	–	–	–	–

Table A.1 (continued) References

B1	Barrett (87b)	F1	Fermon (95)	N2	Nicklin (96)	T4	Tougaard (82)
B2	Barrett (91a)	G1	Giergiel (95,96)	Q1	Quinn (91)	W1	Waldfried (96)
B3	Barrett (91b)	H1	Hayoz (98)	Q2	Quinn (92)	W2	Waldfried (98)
B4	Barrett (92a)	K1	Kalinowski (97)	Q3	Quinn (93)	W3	Weller (85d)
B5	Barrett (93)	K2	Kalinowski (98)	R1	Ritley (98)	W4	White (97)
B6	Blyth (91e)	L1	Li H (92)	S1	Sokolov (89)	W5	Wu RQ (91a,b)
B7	Bylander (94)	L2	Li YS (92)	S2	Stenborg (89a)	Y1	Yang (88)
C1	Chen (92)	M1	Mishra (98a)	T1	Theis-Bröhl (97a)		
D1	Dhesi (95)	M2	Mozley (95)	T2	Theis-Bröhl (97b)		
E1	Eriksson (95)	N1	Nicklin (92)	T3	Tober (96)		

Table A.2 Studies of single–crystal rare–earth metal surfaces (electronic and magnetic structures)

	Electronic Structure	Magnetic Structure	Theory
Sc	B5 B12 O1 O2 P3 T6	–	F6
Y	B1 B2 B3 B4 B5 B9 B12 J4 M9	–	–
La	F3 W9 W10	–	–
Ce	J1 K3 L2 M1 M2 R6 R8 W5 W9 W13	–	D8 E2 H4
Pr	B12 D2 W10	–	–
Nd	C2 C3	C2 C3	–
Sm	C3 L3	C3	H5 R7
Eu	L3	–	R7
Gd	B12 C1 C2 C3 D4 D7 F4 H3 J3 K4 K5 L5 L6 L7 L9 M7 O3 R4 T2 T4 V2 W1 W8	A1 A2 A4 A5 B8 C1 C2 C3 D4 F1 F2 F5 F7 K4 L8 L9 M4 M6 M7 M8 P1 P2 R2 R4 S3 S5 S7 S8 T2 T5 v1 V3 W2 W3 W4 W8	A3 B14 d1 E1 H2 S1 S4 W11
Tb	B10 C2 C3 D5 H6 L1 N1 T4 W12	A4 C2 C3 D6 G1 R3 S6 S7	–
Dy	C2 C3 D5 H1 P4	A4 C2 C3 T3	–
Ho	B11	F8	–
Er	–	B7	–
Tm	B13 D3 L3 N4 N5	–	B13 R7
Yb	B13 J5 L3 T1	–	B13 R7
Lu	O2	–	–
Ln	B5 B6 C2 C3 K1 K2 L3 L4 M3 N2 N3 W6 W7	C2 C3 M5 R1 S2 W6	F9 J2 R5

Table A.2 (continued) References

A1	Ali (93)	D5	Dowben (90)	J5	Joyce (96)	N3	Netzer (86)	T1	Takakuwa (82,84)
A2	André (95)	D6	Dowben (91)	K1	Kaindl (95a)	N4	Nicklin (94)	T2	Tang H (93a,b,c,d)
A3	Anisimov (94)	D7	Dowben (95)	K2	Kaindl (95b)	N5	Nicklin (95)	T3	Theis-Bröhl (96)
A4	Arenholz (95a)	D8	Duo (97)	K3	Kanai (97)	O1	Onsgaard (79)	T4	Thole (93a)
A5	Aspelmeier (94)	E1	Eriksson (95)	K4	Kim (92)	O2	Onsgaard (80a)	T5	Tober (98)
B1	Baptist (88)	E2	Eriksson (97)	K5	Kim (97)	O3	Ortega (94)	T6	Tougaard (81)
B2	Barrett (87a)	F1	Farle (93a,b,94)	L1	LaGraffe (90b)	P1	Pang (94)	v1	van der Laan (96)
B3	Barrett (87b)	F2	Fasolino (93a,b)	L2	Laubschat (92)	P2	Paschen (93)	V2	Vescovo (93a)
B4	Barrett (89a,b,c)	F3	Fedorov (93,94a)	L3	Laubschat (96)	P3	Patchett (94a,b)	V3	Vescovo (93b)
B5	Barrett (92a)	F4	Fedorov (94b,c)	L4	Laubschat (97)	P4	Patthey (93)	W1	Waldfried (96,97)
B6	Barrett (92b)	F5	Fedorov (98)	L5	Li DQ (91a,b)	R1	Rau (82a,94)	W2	Weller (85a,b,c)
B7	Beach (91)	F6	Feibelman (79)	L6	Li DQ (92a,b,93a,94)	R2	Rau (82b,83,86a,87)	W3	Weller (85d)
B8	Berger (94,95)	F7	Fermon (95)	L7	Li DQ (92c)	R3	Rau (86b,88a,b,89a-c,90a)		
B9	Blyth (91a)	F8	Fischer (92)	L8	Li DQ (93b,c)	R4	Reihl (82)	W4	Weller (86a,88)
B10	Blyth (91b)	F9	Freeman (91)	L9	Li DQ (95,96)	R5	Rettori (95)	W5	Weschke (91)
B11	Blyth (91c)	G1	Goedkoop (91)	M1	Mårtensson (82)	R6	Rhee (95)	W6	Weschke (94)
B12	Blyth (92)	H1	Hanyu (96)	M2	Mårtensson (85)	R7	Rosengren (82)	W7	Weschke (95a,b)
B13	Bodenbach (94)	H2	Heinemann (94)	M3	Matthew (95)	R8	Rosina (85,86a)	W8	Weschke (96,97a)
B14	Bylander (94)	H3	Himpsel (83a)	M4	McIlroy (96)	S1	Sabiryanov (97)	W9	Weschke (98)
C1	Carbone (87)	H4	Hjortstam (96)	M5	Miller (93)	S2	Sacchi (91)	W10	Wieliczka (84)
C2	Carbone (90a)	H5	Hong (97)	M6	Mishra (98a)	S3	Salas (90,91,93)	W11	Wu RQ (91a,b)
C3	Carbone (90b)	H6	Hubinger (95)	M7	Mishra (98b)	S4	Skomski (98)	W12	Wu SC (90,91,92)
d1	de Moraes (93)	J1	Jensen (84)	M8	Mulhollan (92)	S5	Starke (93)	W13	Wuilloud (83)
D2	Dhesi (92)	J2	Johansson B (79,80)	M9	Muthe (94)	S6	Starke (94)		
D3	Domke (86)	J3	Jordan (86)	N1	Navas (93)	S7	Starke (95a,b)		
D4	Donath (96)	J4	Jordan (90)	N2	Netzer (82)	S8	Stetter (92a,b)		

Table A.3 Studies of ultra–thin films of rare–earth metals on bcc refractory metal substrates

	Nb	Mo	Ta	W
Sc	–	–	–	L1 Z2
Y	K4 T3	–	T3	–
La	–	L7	S3	B1 L8
Ce	–	T1 K1	R2	G5 I1
Pr	–	–	R1	H1
Nd	–	L6 Z1	–	L6 O2
Sm	d1 S8	N3 S5 S6	S7 T2	C3 L4 M4
Eu	–	–	O1	K2 K3 M6
Gd	K4	L5 M1 M7	–	B2 C1 D2 F1 G1 G2 G6 K3 L2 M1 M3 M8 O2 P1 P2 T4 T5 T6 W1 W2
Tb	–	S2	–	G2 K2 K3 M2 M3 M4 P2
Dy	–	G4 L3 L6 M5 S2	S3	C2 G1 L6
Ho	–	–	–	G3
Er	–	P3 S1 S2	–	–
Tm	–	N1 N2	–	B3
Yb	–	S4 S6	–	–
Lu	–	–	–	–

Table A.3 (continued) References

B1	Bhave (91)	K1	Kamei (96,97)	M6	Melmed (84)	S5	Stenborg (87b,89a)
B2	Bode (98a,c)	K2	Kolaczkiewicz (85)	M7	Mozley (95)	S6	Stenborg (87c)
B3	Burmistrova (83)	K3	Kolaczkiewicz (86,92a,b,93)	M8	Mühlig (98)	S7	Strisland (97)
C1	Ciszewski (84a)	K4	Kwo (86)	N1	Nicklin (92)	S8	Strisland (98)
C2	Ciszewski (84b)	L1	Lamouri (95)	N2	Nicklin (94)	T1	Tanaka (95a,b,96a,b)
C3	Ciszewski (84c)	L2	Li DQ (91b,92a)	N3	Nicklin (96)	T2	Tao (93)
d1	denBoer (88)	L3	Loburets (98)	O1	Olson (95)	T3	Theis-Bröhl (97a)
D2	Dowben (95)	L4	Loginov (92)	O2	Ostertag (97)	T4	Tober (96)
F1	Farle (87,89)	L5	Losovyj (97)	P1	Pascal (97a-d)	T5	Tober (97)
G1	Gonchar (87)	L6	Losovyj (98)	P2	Pascal (98)	T6	Tucker (95,98)
G2	Gonchar (88)	L7	Lozovyi (82)	P3	Player (92)	W1	White (95)
G3	Gonchar (89)	L8	Lozovyi (86)	R1	Raaen (90b)	W2	Wiesendanger (98)
G4	Gonchar (90)	M1	Madey (96)	R2	Raaen (92b)	Z1	Zadorozhnyi (92)
G5	Gu (91,93)	M2	Medvedev (90)	S1	Savaloni (92a,b)	Z2	Zagwijn (97)
G6	Guan (95)	M3	Medvedev (91)	S2	Shakirova (92)		
H1	Hwang (97)	M4	Medvedev (92)	S3	Smereka (95)		
I1	Ivanov (89)	M5	Medvedev (93)	S4	Stenborg (87a,89c)		

Table A.4 Studies of ultra–thin films of rare–earth metals on fcc and hcp metal substrates

	Al	Fe	Ni	Cu	Other
Sc	S10	–	–	–	–
Y	–	–	–	L5	A1 D7
La	–	–	–	–	G5 L6 O2 R3 S7
Ce	B6 R1 R3 R4	K5 K6 W3	O3 O4	B10 O4 R3 R4	B1 B5 B11 G9 H1 H2 H3 K4 R3 S2 S3 T2 T4 W2
Pr	R2	V5	–	–	B11 O1
Nd	–	C2 C3 C4 F6	–	N2 S6	S8
Sm	F1 F3	C3 V2 Z3	R8	A3 F2 F3 J1 J2 T3	D8 F4 G6 K7 P1 R6 V1
Eu	–	–	–	–	A4 B4 B8 M2 S9
Gd	–	C1 C2 C3 C5 K1 L4 M1 P2 P3 T1 V5	D3 D4 L2 L3 P3	D2 D3 D4 D6 L1 L3 R5 S1	G1 G2 T7 Z1 Z2
Tb	–	C2 C3 P3 R7 S4	D5 G4 L3 P3	D5 L3	B7 G3 S5
Dy	–	C2 C3	D3 K2 K3 V5	D1 D2 D3 K3 U1	B2 S2 S3 T8 V3 V4 Y1
Ho	–	–	–	–	–
Er	–	–	–	–	B9 G5 W1
Tm	–	–	–	–	G5
Yb	C6 F5 G7 G8 N3 O5 T5	–	A2 C6 C7 N1 T6	–	B3 G8 O1
Lu	–	–	–	–	B2

Table A.4 (continued) References

A1	Andersen JET (91a,b)	D2	Demri (93b)	H2	Heuberger (93,94)	O2	Ohno (93)	S9	Sørland (93)
A2	Andersen JN (87,88b)	D3	Dottl (93)	H3	Homma (87)	O3	Okane (95)	S10	Strisland (96)
A3	Andersen JN (88a)	D4	Dowben (89)	J1	Jaffey (89)	O4	Okane (96,98)	T1	Taborelli (86a,b)
A4	Arenholz (95b,98a,b)	D5	Dowben (91)	J2	Jørgensen (91)	O5	Onsgaard (84)	T2	Tang J (93a,b)
B1	Baddeley (97)	D6	Dowben (95)	K1	Kachel (92,94)	P1	Pan (96)	T3	Tao (93)
B2	Beach (93)	D7	Du (90)	K2	Kappert (91)	P2	Panaccione (98)	T4	Thromat (96)
B3	Beaurepaire (87,88)	D8	Dubot (90a,b,91,93)	K3	Kappert (93)	P3	Paul (90)	T5	Tibbetts (80)
B4	Berg (92a)	F1	Fäldt (83,84b,86)	K4	Kierren (94a)	R1	Raaen (90a)	T6	Tougaard (90)
B5	Berg (92b,94a)	F2	Fäldt (84a)	K5	Kierren (94b)	R2	Raaen (90b)	T7	Trappmann (97)
B6	Berg (94b)	F3	Fäldt (85)	K6	Kierren (96a,b,c)	R3	Raaen (92a)	T8	Tsui (91)
B7	Berning (98)	F4	Fäldt (88)	K7	Kuriyama (97a,b)	R4	Raaen (92b)	U1	Ufuktepe (93)
B8	Bertran (91,92a,b)	F5	Fasel (97a,b,c)	L1	LaGraffe (89b)	R5	Raiser (92)	V1	Venvik (96)
B9	Borchers (91a,b)	F6	Fermin (98)	L2	LaGraffe (90a)	R6	Rao (95a,98)	V2	Vescovo (92)
B10	Braaten (89)	G1	Gajdzik (95,98)	L3	LaGraffe (91)	R7	Rau (89c,90a)	V3	Vogel (93)
B11	Braaten (91)	G2	Gammon (97)	L4	Landolt (86)	R8	Roe (94)	V4	Vogel (94)
C1	Carbone (87)	G3	Garreau (96)	L5	Liu XM (96)	S1	Scarfe (92)	V5	Vogel (96)
C2	Carbone (90a)	G4	Goedkoop (91)	L6	Lozovyi (84)	S2	Schneider (97)	W1	Waldrop (92)
C3	Carbone (90b)	G5	Gorbatyi (91)	M1	McGrath (96)	S3	Schneider (98)	W2	Warren (93)
C4	Cheng (92)	G6	Gourieux (96)	M2	Melmed (84)	S4	Scholz (94)	W3	Witkowski (97)
C5	Cherief (93)	G7	Greber (89)	N1	Nilsson (87,88)	S5	Schorsch (94)	Y1	Yang (88)
C6	Chorkendorff (85a)	G8	Greber (90)	N2	Nix (87,88,89a,b)	S6	Shaw (92)	Z1	Zhang FP (97)
C7	Chorkendorff (85b)	G9	Gröning (90)	N3	Nyholm (84)	S7	Shikin (94)	Z2	Zhang H (95,96)
D1	Demri (93a)	H1	Hardacre (95)	O1	Öfner (92)	S8	Shimizu (92)	Z3	Zhang XX (94)

ACKNOWLEDGEMENTS

The authors would like to thank all those colleagues who assisted in the production of this book. We are especially grateful to Colin Mason for critical comments during the proof reading of the manuscript and for assistance with the compilation of the tables in the appendix. Adrian Wander is thanked for his comments on the details of LEED calculations. We appreciate the cooperation of those researchers who provided reprints or preprints of their work, and especially electronic copies of figures from published papers, which made the task of preparing the figures for this book considerably easier.

SDB would like to thank the PhD graduates of the Rare Earth Group at the University of Liverpool for the work that they carried out over the years and thus the contributions that they have made to this book — Rob Blyth, Sarnjeet Dhesi, Adam Patchett, Martin Evans, Richard White, Myoung–Ho Lee, Nigel Tucker and Chris Searle — plus the many MSc students, vacation students and visiting students who have spent time with us along the way. It was a pleasure to work with you all.

Dave Norman and the staff of the Synchrotron Radiation Source at Daresbury Laboratory, together with Liverpool staff Andy, Vin and George, are thanked for the continuous support that they have given to our rare–earth photoemission experiments. Dave Fort and Jeff Sutton are thanked for the their willingness to supply the highest quality single crystals of the rare–earth metals that we requested.

SSD would like to thank Francesca for all her support and encouragement. Finally, SDB would like to thank Robin Jordan for introducing him to the rare–earth metals, Tony Begley for working alongside him in those halcyon days, and Hawkwind for providing the soundtrack to countless experiments at the synchrotron.

The following figures have been reproduced (with minor or no modifications) from journals with permission from the publishers:

Fig. 3.7 Reprinted from *Jpn. J. Appl. Phys.* **34** 5774, Tanaka Y, Kamei M and Gotoh Y, Copyright (1995), with permission from the Japanese Journal of Applied Physics.

Fig. 4.6 Reprinted from *Surf. Sci.* **352** 557, Gourieux T, Kierren B, Bertran F, Malterre D and Krill G, Copyright (1996), with permission from Elsevier Science.

Fig. 4.9 Reprinted from *J. Phys.: Condens. Matter* **2** 1335, Du R and Flynn CP, Copyright (1990), with permission from the Institute of Physics Publishing.

Fig. 4.10 Reprinted from *J. Phys.: Condens. Matter* **9** 10615, Waldfried C, McIlroy DN and Dowben PA, Copyright (1997), with permission from the Institute of Physics Publishing Publishing.

Fig. 5.6 Reprinted from *J. Phys.: Condens. Matter* **9** 10615, Waldfried C, McIlroy DN and Dowben PA, Copyright (1997), with permission from the Institute of Physics Publishing.

Fig. 5.10 Reprinted from *Surf. Sci.* **301** 39, Roe GM, de Castilho CMC and Lambert RM, Copyright (1994), with permission from Elsevier Science.

Fig. 5.14 Reprinted from *Europhys. Lett.* **39** 159, Trappmann T, Gajdzik M, Sürgers C and von Löhneysen Hv, Copyright (1997), with permission from the American Institute of Physics.

Fig. 5.15 Reprinted from *Europhys. Lett.* **39** 159, Trappmann T, Gajdzik M, Sürgers C and von Löhneysen Hv, Copyright (1997), with permission from the American Institute of Physics.

Fig. 5.16 Reprinted from *Phys. Rev.* B **57** 3525, Gajdzik I, Trappmann T, Sürgers C and Löhneysen Hv, Copyright (1998), with permission from the American Institute of Physics.

Fig. 5.26 Reprinted from *J. Phys. D: Appl. Phys.* **26** 143, Fuchs G, Mélinon P, Treilleux M and Le Brusq J, Copyright (1993), with permission from the Institute of Physics Publishing.

Fig. 5.27 Reprinted from *J. Magn. Magn. Mat.* **132** 22, Aspelmeier A, Gerhardter F and Baberschke K, Copyright (1994), with permission from Elsevier Science.

Fig. 5.29 Reprinted from *Phys. Rev.* B **51** 1803, Lamouri A, Krainsky IL, Petukhov AG, Lambrecht WRL and Segall B, Copyright (1995), with permission from the American Institute of Physics.

Fig. 5.30 Reprinted from *Surf. Sci.* **245** L163, Bertran F, Gourieux T, Krill G, Alnot M, Ehrhardt JJ and Felsch W, Copyright (1991), with permission from Elsevier Science.

Fig. 5.31 Reprinted from *Surf. Sci.* **301** 39, Roe GM, de Castilho CMC and Lambert RM, Copyright (1994), with permission from Elsevier Science.

Fig. 7.7 Reprinted from *Phys. Rev.* B **46** 9694, Quinn J, Li YS, Jona F and Fort D, Copyright (1992), with permission from the American Institute of Physics.

REFERENCES

Abe Y, Kato A and Nakamura T 1994 *Surf. Int. Anal.* **22** 14

Abell JS 1989 in: *Handbook on the Physics and Chemistry of Rare Earths* eds Gschneidner KA and Eyring L (North–Holland, Amsterdam) Vol.12 p.1

Ali N, Masden JT, Hill P and Chin A 1993 *Int. J. Mod. Phys.* B **7** 496

Allen JW, Johansson LI, Bauer RS, Lindau I and Hagström SBM 1978 *Phys. Rev. Lett.* **41** 1499

Allen JW, Johansson LI, Lindau I and Hagström SBM 1980 *Phys. Rev.* B **21** 1335

Allen JW, Oh SJ, Lindau I, Lawrence JM, Johansson LI and Hagström SBM 1981 *Phys. Rev. Lett.* **46** 1100

Altmann SL and Bradley CJ 1967 *Proc. Phys. Soc. (London)* **92** 764

Alvarado SF, Campagna M and Gudat W 1980 *J. Electron Spectrosc.* **18** 43

Andersen JET and Møller PJ 1991a *Phys. Rev.* B **44** 13645

Andersen JET and Møller PJ 1991b *Surf. Sci.* **258** 247

Andersen JN, Onsgaard J, Nilsson A, Eriksson B, Zdansky E and Mårtensson N 1987 *Surf. Sci.* **189–190** 399

Andersen JN, Chorkendorff I, Onsgaard J, Ghijsen J, Johnson RL and Grey F 1988a *Phys. Rev.* B **37** 4809

Andersen JN, Onsgaard J, Nilsson A, Eriksson B and Mårtensson N 1988b *Surf. Sci.* **202** 183

André G, Aspelmeier A, Schulz B, Farle M and Baberschke K 1995 *Surf. Sci.* **326** 275

Anisimov AV, Borman VD and Popov AP 1994 *Phys. Rev.* B **49** 3874

Arenholz E, Navas E, Starke K, Baumgarten L and Kaindl G 1995a *Phys. Rev.* B **51** 8211

Arenholz E, Starke K and Kaindl G 1995b *J. Electron Spectrosc.* **76** 183

Arenholz E, Starke K, Kaindl G and Jensen PJ 1998a *Phys. Rev. Lett.* **80** 2221

Arenholz E, Starke K, Navas E and Kaindl G 1998b *J. Electron Spectrosc.* **78** 241

Argile C and Rhead GE 1989 *Surf. Sci. Rep.* **10** 277

Arnold CS and Pappas DP 2000 *Phys. Rev. Lett.* **85** 5202

Aspelmeier A, Gerhardter F and Baberschke K 1994 *J. Magn. Magn. Mat.* **132** 22

Baddeley CJ, Stephenson AW, Hardacre C, Tikhov M and Lambert RM 1997 *Phys. Rev.* B **56** 12589

Baer Y and Busch G 1973 *Phys. Rev. Lett.* **31** 35

Baer Y and Schneider WD 1987 in: *Handbook on the Physics and Chemistry of Rare Earths* eds Gschneidner KA, Eyring L and Hüfner S (North–Holland, Amsterdam) Vol.10 p.1
Baptist R, Pellisier A and Chauvet G 1988 *Z. Phys. B – Cond. Matter* **73** 107
Barrett SD and Jordan RG 1987a *Z. Phys. B – Cond. Matter* **66** 375
Barrett SD, Jordan RG and Begley AM 1987b *J. Phys. F: Met. Phys.* **17** L145
Barrett SD, Begley AM, Durham PJ and Jordan RG 1989a *Solid State Commun.* **71** 111
Barrett SD, Jordan RG, Begley AM and Blyth RIR 1989b *Z. Phys. B – Cond. Matter* **76** 137
Barrett SD, Begley AM, Durham PJ, Jordan RG and Temmerman WM 1989c *J. Phys.: Cond. Matter* **1** SB243
Barrett SD, Blyth RIR, Begley AM, Dhesi SS and Jordan RG 1991a *Phys. Rev.* B **43** 4573
Barrett SD, Blyth RIR, Dhesi SS and Newstead K 1991b *J. Phys.: Condens. Matter* **3** 1953
Barrett SD 1992a *Surf. Sci. Rep.* **14** 271
Barrett SD and Blyth RIR 1992b *Z. Phys. B – Cond. Matter* **86** 157
Barrett SD, Dhesi SS, Evans MP and White RG 1993 *Meas. Sci. Technol.* **4** 114
Bastl Z, Cerny S and Kovár M 1993 *Appl. Surf. Sci.* **68** 275
Bauer E 1977 *J. Physique* **38** 146
Bauer E 1982 *Appl. Surf. Sci.* **11–12** 479
Bauer E and van der Merwe JH 1986 *Phys. Rev.* B **33** 3657
Beach RS, Borchers JA, Erwin RW, Rhyne JJ, Matheny A, Flynn CP and Salamon MB 1991 *J. Appl. Phys.* **69** 4535
Beach RS, Matheny A, Salamon MB, Flynn CP, Borchers JA, Erwin RW and Rhyne JJ 1993 *J. Appl. Phys.* **73** 6901
Beaudry BJ and Gschneidner KA 1978a in: *Handbook on the Physics and Chemistry of Rare Earths* eds Gschneidner KA and Eyring L (North–Holland, Amsterdam) Vol.1 p.173
Beaudry BJ and Gschneidner KA 1978b in: *Handbook on the Physics and Chemistry of Rare Earths* eds Gschneidner KA and Eyring L (North–Holland, Amsterdam) Vol.1 p.224
Beaurepaire E, Brouder C, Carriere B, Chandesris D, Krill G, Lecante J and Legare P 1987 *J. Physique* **48** 939
Beaurepaire E, Carriere B, Chandesris D, Brouder C, Krill G, Legare P and Lecante J 1988 *J. Physique* **49** 1719
Becker JJ 1970 *J. Appl. Phys.* **41** 1055
Beeby J 1968 *J. Phys. C: Solid State Phys.* **1** 82
Begley AM 1990 PhD thesis, University of Birmingham, UK
Begley AM, Jordan RG, Temmerman WM and Durham PJ 1990 *Phys. Rev.* B **41** 11780
Berg C, Raaen S and Ruckman MW 1992a *J. Phys.: Condens. Matter* **4** 4213
Berg C and Raaen S 1992b *J. Phys.: Condens. Matter* **4** 8021
Berg C, Raaen S, Borg A and Venvik HJ 1994a *Phys. Rev.* B **50** 1976
Berg C, Raaen S and Borg A 1994b *Surf. Sci.* **303** 114
Berger A, Pang AW and Hopster H 1994 *J. Magn. Magn. Mat.* **137** L1
Berger A, Pang AW and Hopster H 1995 *Phys. Rev.* B **52** 1078
Berning GLP 1976 *Surf. Sci.* **61** 673

Berning GLP, Alldredge GP and Viljoen PE 1981 *Surf. Sci.* **104** L225

Berning GLP and Swart HC 1998 *Surf. Int. Anal.* **26** 420

Bertel E, Netzer FP and Matthew JAD 1981 *Surf. Sci.* **103** 1

Bertel E, Strasser G, Netzer FP and Matthew JAD 1982 *Phys. Rev.* B **25** 3374

Bertran F, Gourieux T, Krill G, Alnot M, Ehrhardt JJ and Felsch W 1991 *Surf. Sci.* **245** L163

Bertran F, Gourieux T, Krill G, Ravet-Krill MF, Alnot M, Ehrhardt JJ and Felsch W 1992a *Phys. Rev.* B **46** 7829

Bertran F, Gourieux T, Krill G, Alnot M, Ehrhardt JJ and Felsch W 1992b *Surf. Sci.* **270** 731

Besenbacher F 1996 *Rep. Prog. Phys.* **59** 1737

Bhave AS and Kanitkar PL 1991 *J. Phys. D: Appl. Phys.* **24** 454

Biberian JP and Somorjai GA 1979 *J. Vac. Sci. Technol.* **16** 2973

Binnig G, Rohrer H, Gerber C and Weibel E 1982 *Phys. Rev. Lett.* **49** 57

Björneholm O, Andersen JN, Christiansen M, Nilsson A, Wigren C, Onsgaard J, Stenborg A and Mårtensson N 1989 *J. Phys.: Condens. Matter* **1** SB271

Blaha P, Schwarz K and Dederichs PH 1988 *Phys. Rev.* B **38** 9368

Blakemore J S 1988 *Solid State Physics* (Cambridge University, Cambridge)

Blyth RIR, Andrews PT and Barrett SD 1991a *J. Phys.: Condens. Matter* **3** 2827

Blyth RIR, Dhesi SS, Patchett AJ, Mitrelias T, Prince NP and Barrett SD 1991b *J. Phys.: Condens. Matter* **3** 6165

Blyth RIR, Barrett SD, Dhesi SS, Cosso R, Heritage N, Begley AM and Jordan RG 1991c *Phys. Rev.* B **44** 5423

Blyth RIR, Patchett AJ, Dhesi SS, Cosso R and Barrett SD 1991d *J. Phys.: Condens. Matter* **3** S287

Blyth RIR, Cosso R, Dhesi SS, Newstead K, Begley AM, Jordan RG and Barrett SD 1991e *Surf. Sci.* **251–252** 722

Blyth RIR, Dhesi SS, Gravil PA, Newstead K, Cosso R, Cole RJ, Patchett AJ, Mitrelias T, Prince NP and Barrett SD 1992 *J. Alloys Compounds* **180** 259

Bode M, Getzlaff M, Heinze S, Pascal R and Wiesendanger R 1998a *Appl. Phys.* A **66** S121

Bode M, Pascal R, Getzlaff M and Wiesendanger R 1998b *Acta Phys. Pol.* A **93** 273

Bode M, Getzlaff M and Wiesendanger R 1998c *Phys. Rev. Lett.* **81** 4256

Bodenbach M, Höhr A, Laubschat C, Kaindl G and Methfessel M 1994 *Phys. Rev.* B **50** 14446

Bohr J, Gibbs D, Axe JD, Moncton DE, D'Amico KL, Majkrzak CF, Kwo J, Hong M, Chien CL and Jensen J 1989 *Physica* B **159** 93

Bondarenko BV and Makhov VI 1971 *Soviet Phys. – Solid State* **12** 1646

Borchers JA, Salamon MB, Erwin RW, Rhyne JJ, Du RR and Flynn CP 1991a *Phys. Rev.* B **43** 3123

Borchers JA, Salamon MB, Erwin RW, Rhyne JJ, Nieuwenhuys GJ, Du RR, Flynn CP and Beach RS 1991b *Phys. Rev.* B **44** 11814

Braaten NA, Grepstad JK and Raaen S 1989 *Phys. Rev.* B **40** 7969

Braaten NA and Raaen S 1991 *Phys. Scr.* **43** 430

Bracconi P, Pörschke E and Lässer R 1988 *Appl. Surf. Sci.* **32** 392

Bridge ME, Comrie CM and Lambert RM 1977 *Surf. Sci.* **67** 393

Bridgman PW 1927 *Proc. Am. Acad. Arts Sci.* **62** 207

Bruce LA and Jaeger H 1978 *Phil. Mag.* **38** 223

Brundle CR and Baker AD (eds) 1977 *Electron Spectroscopy: Volume 3* (Academic, New York)

Brune H 1998 *Surf. Sci. Rep.* **31** 121

Bucher E, Schmidt PH, Jayaraman A, Andres K, Maita JP, Nassau K and Dernier PD 1970 *Phys. Rev.* B **2** 3911

Burmistrova OP, Mitsev MA and Mukhuchev AM 1983 *Fiz. Tverd. Tela* **25** 47

Bylander DM and Kleinman L 1994 *Phys. Rev.* B **50** 4996

Cadieu FJ and Douglass DH 1969 *J. Appl. Phys.* **40** 2376

Campagna M, Wertheim GK and Baer Y 1979 in: *Photoemission in Solids II* eds Ley L and Cardona M (Springer–Verlag, Berlin) ch.4

Carbone C and Kisker E 1987 *Phys. Rev.* B **36** 1280

Carbone C, Rochow R, Braicovich L, Jungblut R, Kachel T, Tillman D and Kisker E 1990a *Phys. Rev.* B **41** 3866

Carbone C, Rochow R, Braicovich L, Jungblut R, Kachel T, Tillmann D and Kisker E 1990b *Vacuum* **41** 496

Cardona M and Ley L (eds) 1978 in: *Photoemission in Solids I* (Springer–Verlag, Berlin) ch.1

Cardona M and Ley L (eds) 1978 *Photoemission in Solids I* (Springer–Verlag, Berlin) appx

Carlson ON, Schmidt FA and Peterson DT 1966 *J. Less-Common Metals* **19** 1

Carnahan TG and Scott TE 1973 *Metall. Trans.* **4** 27

Cerri A, Mauri D and Landolt M 1983 *Phys. Rev.* B **27** 6526

Che H, Lin DL, Xia Y, Zheng H and Li HX 1992 *Phys. Rev.* B **46** 13501

Chen SP 1992 *Surf. Sci.* **264** L162

Cheng YT and Chen YL 1992 *Appl. Phys. Lett.* **60** 1951

Cherief N, Givord D, Liénard A, Mackay K, McGrath OFK, Rebouillat JP, Robaut F and Souche Y 1993 *J. Magn. Magn. Mat.* **121** 94

Chorkendorff I, Onsgaard J, Aksela H and Aksela S 1983 *Phys. Rev.* B **27** 945

Chorkendorff I, Kofoed J and Onsgaard J 1985a *Surf. Sci.* **152–153** 749

Chorkendorff I, Onsgaard J, Schmidt-May J and Nyholm R 1985b *Surf. Sci.* **160** 587

Ciszewski A and Melmed AJ 1984a *Surf. Sci.* **145** L471

Ciszewski A and Melmed AJ 1984b *J. Cryst. Growth* **69** 253

Ciszewski A and Melmed AJ 1984c *J. Physique* **45** 39

Coey JMD, Skumryev V and Gallagher K 1999 *Nature* **401** 35

Cox PA, Baer Y and Jørgensen CK 1973 *Chem. Phys. Lett.* **22** 433

Cox PA, Lang JK and Baer Y 1981 *J. Phys. F: Metal Phys.* **11** 113

Das SG 1976 *Phys. Rev.* B **13** 3978

Davis HL and Zehner DM 1980 *J. Vac. Sci. Technol.* **17** 190

Davis HL, Hannon JB, Ray KB and Plummer EW 1992 *Phys. Rev. Lett.* **68** 2632

Davisson CJ and Germer LH 1927 *Phys. Rev.* **30** 705

de Asha AM and Nix RM 1995 *Surf. Sci.* **322** 41

de Moraes JNB and Figueiredo W 1993 *Phys. Rev.* B **47** 5819

Demri B, Burggraf C and Mesboua N 1993a *Appl. Surf. Sci.* **65–66** 59

Demri B, Burggraf C and Mesboua N 1993b *Nucl. Instr. Meth. Phys. Res.* B **79** 450

denBoer ML, Chang CL, Horn S and Murgai V 1988 *Phys. Rev.* B **37** 6605

Desjonquères MC, Spanjaard D, Lassailly Y and Guillot C 1980 *Solid State Commun.* **34** 807

Dhesi SS, Blyth RIR, Cole RJ, Gravil PA and Barrett SD 1992 *J. Phys.: Condens. Matter* **4** 9811

Dhesi SS, White RG, Patchett AJ, Evans MP, Lee MH, Blyth RIR, Leibsle FM and Barrett SD 1995 *Phys. Rev.* B **51** 17946

Dimmock JO and Freeman AJ 1964 *Phys. Rev. Lett.* **13** 750

Dodson BW, Myers DR, Datye AK, Kaushik VS, Kendall DL and Martinez-Tovar B 1988 *Phys. Rev. Lett.* **61** 2681

Domke M, Laubschat C, Prietsch M, Mandel T, Kaindl G and Schneider WD 1986 *Phys. Rev. Lett.* **56** 1287

Donath M, Gubanka B and Passek F 1996 *Phys. Rev. Lett.* **77** 5138

Dottl L, Onellion M, Li DQ and Dowben PA 1993 *Z. Phys. B – Cond. Matter* **90** 93

Dowben PA, LaGraffe D and Onellion M 1989 *J. Phys.: Condens. Matter* **1** 6571

Dowben PA, LaGraffe D, Li DQ, Dottl L, Hwang C, Ufuktepe Y and Onellion M 1990 *J. Phys.: Condens. Matter* **2** 8801

Dowben PA, LaGraffe D, Li DQ, Miller A, Zhang L, Dottl L and Onellion M 1991 *Phys. Rev.* B **43** 3171

Dowben PA, Li DQ, Zhang JD and Onellion M 1995 *J. Vac. Sci. Technol.* A **13** 1549

Dowben PA, McIlroy DN and Li DQ 1997 in: *Handbook on the Physics and Chemistry of Rare Earths* eds Gschneidner KA and Eyring L (North–Holland, Amsterdam) Vol.24 p.1

Du R, Tsui F and Flynn CP 1988 *Phys. Rev.* B **38** 2941

Du R and Flynn CP 1990 *J. Phys.: Condens. Matter* **2** 1335

Dubot P, Alleno E, Barthés-Labrousse MG, Binns C and Norris C 1990a *Vacuum* **41** 753

Dubot P, Barthés-Labrousse MG and Langeron JP 1990b *Surf. Int. Anal.* **16** 188

Dubot P, Denjean P and Barthés-Labrousse MG 1991 *Surf. Sci.* **258** 35

Dubot P, Alleno E, Barthés-Labrousse MG, Binns C, Nicklin CL, Norris C and Ravot D 1993 *Surf. Sci.* **282** 1

Duc TM, Guillot C, Lassailly Y, Lecante J, Jugnet Y and Vedrine JC 1979 *Phys. Rev. Lett.* **43** 789

Duo L 1997 *Surf. Sci.* **377** 160

Duthie JC and Pettifor DG 1977 *Phys. Rev. Lett.* **38** 564

Edwards LR and Legvold S 1968 *Phys. Rev.* **176** 753

Egelhoff WF and Tibbetts GG 1980 *Phys. Rev. Lett.* **44** 482

Egelhoff WF 1987 *Surf. Sci. Rep.* **6** 253

Erbudak M, Kalt P, Schlapbach L and Bennemann K 1983 *Surf. Sci.* **126** 101

Eriksson O, Albers RC, Boring AM, Fernando GW, Hao YG and Cooper BR 1991 *Phys. Rev.* B **43** 3137

Eriksson O, Ahuja R, Ormeci A, Trygg J, Hjortstam O, Söderlind P, Johansson B and Wills JM 1995 *Phys. Rev.* B **52** 4420

Eriksson O, Trygg J, Hjortstam O, Johansson B and Wills JM 1997 *Surf. Sci.* **382** 93

Erskine JL and Stern EA 1973 *Phys. Rev.* B **8** 1239

Fadley CS 1990 in: *Synchrotron Radiation Research: Advances in Surface Science* ed. Bachrach R Z (Plenum, New York) Vol.1

Fäldt Å and Myers HP 1983 *Solid State Commun.* **48** 253

Fäldt Å and Myers HP 1984a *Phys. Rev. Lett.* **52** 1315

Fäldt Å and Myers HP 1984b *Phys. Rev.* B **30** 5481

Fäldt Å and Myers HP 1985 *J. Magn. Magn. Mat.* **47–48** 225

Fäldt Å and Myers HP 1986 *Phys. Rev.* B **34** 6675

Fäldt Å, Kristensson DK and Myers HP 1988 *Phys. Rev.* B **37** 2682

Farle M, Zomack M and Barberschke K 1985 *Surf. Sci.* **160** 205

Farle M and Baberschke K 1987 *Phys. Rev. Lett.* **58** 511

Farle M, Berghaus A and Baberschke K 1989 *Phys. Rev.* B **39** 4838

Farle M, Baberschke K, Stetter U, Aspelmeier A and Gerhardter F 1993a *Phys. Rev.* B **47** 11571

Farle M, Lewis WA and Baberschke K 1993b *Appl. Phys. Lett.* **62** 2728

Farle M and Lewis WA 1994 *J. Appl. Phys.* **75** 5604

Fasel R, Gierer M, Bludau H, Aebi P, Osterwalder J and Schlapbach L 1997a *Surf. Sci.* **374** 104

Fasel R, Aebi P, Osterwalder J and Schlapbach L 1997b *Surf. Sci.* **394** 129

Fasel R, Aebi P and Schlapbach L 1997c *Surf. Rev. Lett.* **4** 1155

Fasolino A, Carra P and Altarelli M 1993a *Phys. Rev.* B **47** 3877

Fasolino A, Carra P and Altarelli M 1993b *J. Magn. Magn. Mat.* **121** 194

Faulkner JS 1982 *Prog. Mat. Sci.* **27** 1

Fedorov AV, Laubschat C, Starke K, Weschke E, Barholz KU and Kaindl G 1993 *Phys. Rev. Lett.* **70** 1719

Fedorov AV, Höhr A, Weschke E, Starke K, Adamchuk VK and Kaindl G 1994a *Phys. Rev.* B **49** 5117

Fedorov AV, Arenholz E, Starke K, Navas E, Baumgarten L, Laubschat C and Kaindl G 1994b *Phys. Rev. Lett.* **73** 601

Fedorov AV, Starke K and Kaindl G 1994c *Phys. Rev.* B **50** 2739

Fedorov AV, Valla T, Huang DJ, Reisfeld G, Loeb F, Liu F and Johnson PD 1998 *J. Electron Spectrosc.* **92** 19

Feibelman PJ and Hamann DR 1979 *Solid State Commun.* **31** 413

Feibelman PJ 1983 *Phys. Rev.* B **27** 2531

Fermin JR, Azevedo A, Rezende SM, Pereira LG and Teixeira S 1998 *J. Appl. Phys.* **83** 4869

Fermon C, McGrath OFK and Givord D 1995 *Physics* B **213** 236

Fernando GW, Cooper BR, Ramana MV, Krakauer H and Ma CQ 1986 *Phys. Rev. Lett.* **56** 2299

Fischer P, Schütz G, Scherle S, Knülle M, Stähler S and Wiesinger G 1992 *Solid State Commun.* **82** 857

Fleming GS and Loucks TL 1968a *Phys. Rev.* **173** 685

Fleming GS, Liu SH and Loucks TL 1968b *Phys. Rev. Lett.* **21** 1524

Fleming GS, Liu SH and Loucks TL 1969 *J. Appl. Phys.* **40** 1285

Flynn CP and Tsui F 1993 in: *Magnetism and Structure in Systems of Reduced Dimension* eds Farrow RFC et al (Plenum, New York) p.195

Fort D 1987 *J. Less-Common Metals* **134** 45

Fort D 1989 *J. Cryst. Growth* **94** 85

Fort D 1991 *J. Alloys Compounds* **177** 31

Freeman AJ, Dimmock JO and Watson RE 1966 *Phys. Rev. Lett.* **16** 94

Freeman AJ and Wu RQ 1991 *Prog. Theor. Phys. Suppl.* **106** 397

Friedman DJ, Carbone C, Bertness KA and Lindau I 1986 *J. Electron Spectrosc.* **41** 59

Fruchart O, Jaren S and Rothman J 1998 *Appl. Surf. Sci.* **135** 218

Fuchs G, Mélinon P, Treilleux M and Le Brusq J 1993 *J. Phys. D: Appl. Phys.* **26** 143

Fukuda Y, Lancaster GM, Honda F and Rabalais JW 1978 *Phys. Rev.* B **18** 6191

Gajdzik I, Paschen U, Sürgers C and Löhneysen Hv 1995 *Z. Phys. B – Cond. Matter* **98** 541

Gajdzik I, Trappmann T, Sürgers C and Löhneysen Hv 1998 *Phys. Rev.* B **57** 3525

Gammon WJ, Mishra SR, Pappas DP, Goodman KW, Tobin JG, Schumann FO, Willis R, Denlinger JD, Rotenberg E, Warwick A and Smith NV 1997 *J. Vac. Sci. Technol.* A **15** 1755

Garreau G, Schorsch V, Beaurepaire E, Scheurer F, Carriere B and Farle M 1996 *J. Magn. Magn. Mat.* **156** 81

Gasgnier M 1995 in: *Handbook on the Physics and Chemistry of Rare Earths* eds Gschneidner KA and Eyring L (North–Holland, Amsterdam) Vol.20 p.105

Gerken F, Barth J and Kunz C 1981 *Phys. Rev. Lett.* **47** 993

Gerken F, Barth J, Kammerer R, Johansson LI and Flodström A 1982 *Surf. Sci.* **117** 468

Gerken F, Flodström AS, Barth J, Johansson LI and Kunz C 1985 *Phys. Scr.* **32** 43

Getzlaff M, Ostertag C, Fecher GH, Cherepkov NA and Schönhense G 1994 *Phys. Rev. Lett.* **73** 3030

Getzlaff M, Paul J, Bansmann J, Ostertag C, Fecher GH and Schönhense G 1996 *Surf. Sci.* **352** 123

Getzlaff M, Bode M and Wiesendanger R 1998a *Surf. Sci.* **410** 189

Getzlaff M, Bode M, Heinze S, Pascal R and Wiesendanger R 1998b *J. Magn. Magn. Mat.* **184** 155

Giergiel J, Pang AW, Hopster H, Guo X, Tong SY and Weller D 1995 *Phys. Rev.* B **51** 10201

Giergiel J, Pang AW, Hopster H, Guo X, Tong SY and Weller D 1996 *Phys. Rev.* B **54** 17223

Ginatempo B, Durham PJ and Gyorffy BL 1989 *J. Phys.: Condens. Matter* **1** 6483

Glötzel D and Fritsche L 1977 *Phys. Stat. Sol.* B **79** 85

Glötzel D 1978 *J. Phys. F: Metal Phys.* **8** L163

Goedkoop JB, Grioni M and Fuggle JC 1991 *Phys. Rev.* B **43** 1179

Gonchar FM, Medvedev VK, Smereka TP, Losovyj YB and Babkin GV 1987 *Fiz. Tverd. Tela* **29** 2833; *Sov. Phys. Solid State* **29** 1629

Gonchar FM, Smereka TP, Stepanovskii SI and Babkin GV 1988 *Fiz. Tverd. Tela* **30** 3541

Gonchar FM, Medvedev VK, Smereka TP and Savichev VV 1989 *Fiz. Tverd. Tela* **31** 249

Gonchar FM, Medvedev VK, Smereka TP and Babkin GV 1990 *Fiz. Tverd. Tela* **32** 1872

Gorbatyi NA, Achilov AU, Pulatova S, Reshetnikova LV and Mutavadg N 1991 *Iz. Akad. Nauk SSSR Ser. Fiz.* **55** 2381

Gotoh Y and Fukuda H 1989 *Surf. Sci.* **223** 315

Gourieux T, Kierren B, Bertran F, Malterre D and Krill G 1996 *Surf. Sci.* **352** 557

Greber T, Osterwalder J and Schlapbach L 1989 *Phys. Rev.* B **40** 9948

Greber T, Büchler S and Schlapbach L 1990 *Vacuum* **41** 556

Grill L, Ramsey MG, Matthew JAD and Netzer FP 1997 *Surf. Sci.* **380** 324

Gröning P, Greber T, Osterwalder J and Schlapbach L 1990 *Vacuum* **41** 1439

Grosshans WA, Vohra YK and Holzapfel WB 1982 *J. Magn. Magn. Mat.* **29** 282

Gu C, Wu X, Olson CG and Lynch DW 1991 *Phys. Rev. Lett.* **67** 1622

Gu C, Olson CG and Lynch DW 1993 *Phys. Rev.* B **48** 12178

Guan J, Campbell RA and Madey TE 1995 *Surf. Sci.* **341** 311

Gupta SC and Gupta YM 1985 *J. Appl. Phys.* **57** 2464

Gustafsson DR, McNutt JD and Roellig LO 1969 *Phys. Rev.* **183** 435

Hagström SBM, Nordling C and Siegbahn K 1964 *Z. Phys.* **178** 433

Hagström SBM, Hedén PO and Löfgren H 1970 *Solid State Commun.* **8** 1245

Hannon JB and Plummer EW 1991 *Abstracts of 12th Europ. Conf. on Surf. Sci.* **15F** 96

Hanyu T, Ishii H, Hashimoto S, Yokoyama T, Jokura K, Sato H and Miyahara T 1996 *J. Electron Spectrosc.* **78** 67

Hardacre C, Roe GM and Lambert RM 1995 *Surf. Sci.* **326** 1

Harima H, Kobayashi N, Takegahara K and Kasuya T 1985 *J. Magn. Magn. Mat.* **52** 367

Harmon BN and Freeman AJ 1974 *Phys. Rev.* B **10** 1979

Harmon BN 1979 *J. Physique* **40** 65

Hayoz J, Sarbach S, Pillo T, Boschung E, Naumovic D, Aebi P and Schlapbach L 1998 *Phys. Rev.* B **58** R4270

Hedén PO, Löfgren H and Hagström SBM 1971 *Phys. Rev. Lett.* 26 432

Hedén PO, Löfgren H and Hagström SBM 1972 *Phys. Stat. Sol.* 49 721

Heinemann M and Temmerman WM 1994 *Surf. Sci.* **309** 1121

Heinz K 1995 *Rep. Prog. Phys.* **58** 637

Helgesen G, Tanaka Y, Hill JP, Wochner P, Gibbs D, Flynn CP and Salamon MB 1997 *Phys. Rev.* B **56** 2635

Heuberger M, Dietler G, Nowak S and Schlapbach L 1993 *J. Vac. Sci. Technol.* A **11** 2707

Heuberger M, Dietler G and Schlapbach L 1994 *Surf. Sci.* **314** 13

Hill HH and Kmetko EA 1975 *J. Phys. F: Metal Phys.* **5** 1119

Himpsel FJ and Reihl B 1983a *Phys. Rev.* B **28** 574

Himpsel FJ 1983b *Adv. Phys.* **32** 1

Himpsel FJ 1990 *Surf. Sci. Rep.* **12** 1

Hjortstam O, Trygg J, Johansson B, Eriksson O and Wills JM 1996 *J. Appl. Phys.* **79** 5837

Hocking WH and Matthew JAD 1990 *J. Phys.: Condens. Matter* **2** 3643

Homma H, Yang KY and Schuller IK 1987 *Phys. Rev.* B **36** 9435

Hong SC, Kwon YS, Rho TH, Lee MS and Illee J 1997 *J. Korean Phys. Soc.* **30** 87

Hopkinson JFL, Pendry JB and Titterington DJ 1980 *Computer Phys. Commun.* **19** 69

Hsu CC, Ho J, Qian JJ and Wang YT 1991 *J. Vac. Sci. Technol.* A **9** 998

Huang YS and Murgai V 1989 *Solid State Commun.* **69** 873

Huang JCA, Du RR and Flynn CP 1991 *Phys. Rev.* B **44** 4060

Hubinger F, Schüssler-Langeheine C, Fedorov AV, Starke K, Weschke E, Höhr A, Vandré S and Kaindl G 1995 *J. Electron Spectrosc.* **76** 535

Hüfner S 1992 *Z. Phys. B – Cond. Matter* **86** 241

Hwang C 1997 *Surf. Sci.* **385** 328

Ionov AM, Volkov VT and Nikiforova TV 1995 *J. Alloys Compounds* **223** 91

Irvine SJC, Young RC, Fort D and Jones DW 1978 *J. Phys. F: Metal Phys.* **8** L269

Ivanov VA, Kirsanova TS and Tumarova TA 1989 *Fiz. Tverd. Tela* **31** 82

Jackson C 1969 *Phys. Rev.* **178** 949

Jaffey DM, Gellman AJ and Lambert RM 1989 *Surf. Sci.* **214** 407

Jang PW, Wang D and Doyle WD 1997 *J. Appl. Phys.* **81** 4664

Jenkins AC and Temmerman WM 1999 *J. Magn. Magn. Mat.* **198–199** 567

Jensen E and Wieliczka DM 1984 *Phys. Rev.* B **30** 7340

Jensen PJ, Dreyssé H and Bennemann KH 1992 *Surf. Sci.* **270** 627

Jepsen DW 1980 *Phys. Rev.* B **22** 5701

Jiang Q and Wang GC 1995 *Surf. Sci.* **324** 357

Johansen G and Mackintosh AR 1970 *Solid State Commun.* **8** 121

Johansson B 1974 *Phil. Mag.* **30** 469

Johansson B 1979 *Phys. Rev.* B **19** 6615

Johansson B and Mårtensson N 1980 *Phys. Rev.* B **21** 4427

Johansson B and Mårtensson N 1987 in: *Handbook on the Physics and Chemistry of Rare Earths* eds Gschneidner KA, Eyring L and Hüfner S (North–Holland, Amsterdam) Vol.10 p.361

Johansson B, Nordström L, Eriksson O and Brooks MSS 1991 *Phys. Scr.* **T39** 100

Johansson LI, Allen JW, Gustafsson T, Lindau I and Hagström SBM 1978 *Solid State Commun.* **28** 53

Johansson LI, Allen JW, Lindau I, Hecht MH and Hagström SBM 1980 *Phys. Rev.* B **21** 1408

Johansson LI, Allen JW and Lindau I 1981 *Phys. Lett.* **86A** 442

Johansson LI, Flodström A, Hörnström SE, Johansson B, Barth J and Gerken F 1982 *Surf. Sci.* **117** 475

Jona F, Strozier JA and Marcus PM 1985 in: *The Structure of Surfaces* eds Van Hove MA and Tong SY (Springer, Berlin) p.92

Jordan RG 1974 *Contemp. Phys.* **15** 375

Jordan RG, Jones DW and Hall MG 1974a *J. Cryst. Growth* **24–25** 568

Jordan RG, Jones DW and Mattocks PG 1974b *J. Less-Common Metals* **34** 25

Jordan RG 1986 *Phys. Scr.* T**13** 22

Jordan RG, Begley AM, Barrett SD, Durham PJ and Temmerman WM 1990 *Solid State Commun.* **76** 579

Jørgensen B, Christiansen M and Onsgaard J 1991 *Surf. Sci.* **251** 519

Joyce JJ, Andrews AB, Arko AJ, Bartlett RJ, Blyth RIR, Olson CG, Benning PJ, Canfield PC and Poirier DM 1996 *Phys. Rev.* B **54** 17515

Kachel T, Rochow R, Gudat W, Jungblut R, Rader O and Carbone C 1992 *Phys. Rev.* B **45** 7267

Kachel T, Gudat W and Holldack K 1994 *Appl. Phys. Lett.* **64** 655

Kahn IH 1975 *Surf. Sci.* **48** 537

Kaindl G, Schneider WD, Laubschat C, Reihl B and Mårtensson N 1983 *Surf. Sci.* **126** 105

Kaindl G, Weschke E, Laubschat C, Ecker R and Höhr A 1993 *Physica* B **188** 44

Kaindl G, Höhr A, Weschke E, Vandré S, Schüssler-Langeheine C and Laubschat C 1995a *Phys. Rev.* B **51** 7920

Kaindl G 1995b *J. Alloys Compounds* **223** 265

Kalff M, Comsa G and Michely T 1998 *Phys. Rev. Lett.* **81** 1255

Kalinowski R, Baczewski LT, Wawro A, Meyer C and Rauluszkiewicz J 1997 *Acta Phys. Pol.* A **93** 409

Kalinowski R, Baczewski LT, Baran M, Givord D, Meyer C and Rauluszkiewicz J 1998 *Appl. Phys.* A **66** S1205

Kamei M, Tanaka Y and Gotoh Y 1996 *Jpn. J. Appl. Phys.* **35** L164

Kamei M, Mizoguchi Y and Gotoh Y 1997 *Jpn. J. Appl. Phys.* **36** 829

Kammerer R, Barth J, Gerken F, Flodström A and Johansson LI 1982 *Solid State Commun.* **41** 435

Kanai K, Tezuka Y, Fujisawa M, Harada Y, Shin S, Schmerber G, Kappler JP, Parlebas JC and Kotani A 1997 *Phys. Rev.* B **55** 2623

Kaneyoshi T 1991 *J. Phys.: Condens. Matter* **3** 4497

Kappert RJH, Sacchi M, Goedkoop JB, Grioni M and Fuggle JC 1991 *Surf. Sci.* **248** L245

Kappert RJH, Vogel J, Sacchi M and Fuggle JC 1993 *Phys. Rev.* B **48** 2711

Kayser FX 1970 *Phys. Rev. Lett.* **25** 662

Keeton SC and Loucks TL 1968 *Phys. Rev.* **168** 672

Kierren B, Gourieux T, Bertran F and Krill G 1994a *Phys. Rev.* B **49** 1976

Kierren B, Bertran F , Gourieux T and Krill G 1994b *J. Phys.: Condens. Matter* **6** L201

Kierren B, Bertran F, Gourieux T, Witkowski N, Malterre D and Krill G 1996a *Phys. Rev.* B **53** 5015

Kierren B, Bertran F, Gourieux T, Malterre D and Krill G 1996b *Europhys. Lett.* **33** 35

Kierren B, Bertran F, Witkowski N, Gourieux T, Malterre D, Finazzi M, Hricovini K and Krill G 1996c *Surf. Sci.* **352** 817

Kim B, Andrews AB, Erskine JL, Kim KJ and Harmon BN 1992 *Phys. Rev. Lett.* **68** 1931

Kim BS and Kim KJ 1997 *J. Korean Phys. Soc.* **30** 83

Kittel C 1996 *Introduction to Solid State Physics* (Wiley, Chichester)

Kmetko EA 1971 in: *Electronic Density of States* ed. Bennett LH (NBS Spec. Publ. 323) p.233

Kolaczkiewicz J and Bauer E 1985 *Surf. Sci.* **154** 357

Kolaczkiewicz J and Bauer E 1986 *Surf. Sci.* **175** 487

Kolaczkiewicz J and Bauer E 1992a *Surf. Sci.* **265** 39

Kolaczkiewicz J and Bauer E 1992b *Surf. Sci.* **273** 109

Kolaczkiewicz J and Bauer E 1993 *Phys. Rev.* B **47** 16506

Koskenmaki DC and Gschneidner KA 1978 in: *Handbook on the Physics and Chemistry of Rare Earths* eds Gschneidner KA and Eyring L (North–Holland, Amsterdam) Vol.1 p.337

Kremeyer S, Speidel KH, Busch H, Grabowy U, Knopp U, Cub J, Bussas M, Maier-Komor P, Gerber J and Meens A 1993 *Hyperfine Int.* **78** 235

Krutzen BCH and Springelkamp F 1989 *J. Phys.: Condens. Matter* **1** 8369

Kunz C 1979 in: *Photoemission in Solids II* eds Ley L and Cardona M (Springer–Verlag, Berlin) ch.6

Kunze U and Kowalsky W 1988 *Appl. Phys. Lett.* **53** 367

Kuriyama T, Kunimori K and Nozoye H 1997a *Appl. Surf. Sci.* **121** 575

Kuriyama T, Kunimori K and Nozoye H 1997b *J. Phys. Chem.* B **101** 11172

Kwo J, Hong M and Nakahara S 1986 *Appl. Phys. Lett.* **49** 319

Lademan WJ, See AK, Klebanoff LE and van der Laan G 1996 *Phys. Rev.* B **54** 17191

LaGraffe D, Dowben PA and Onellion M 1989a *Phys. Rev.* B **40** 970

LaGraffe D, Dowben PA and Onellion M 1989b *Phys. Rev.* B **40** 3348

LaGraffe D, Dowben PA and Onellion M 1989c *Mat. Res. Soc. Symp. Proc.* **151** 71

LaGraffe D, Dowben PA and Onellion M 1990a *J. Vac. Sci. Technol.* A **8** 2738

LaGraffe D, Dowben PA and Onellion M 1990b *Phys. Lett.* A **147** 240

LaGraffe D, Dowben PA, Dottl L, Ufuktepe Y and Onellion M 1991 *Z. Phys. B – Cond. Matter* **82** 47

Lamouri A, Krainsky IL, Petukhov AG, Lambrecht WRL and Segall B 1995 *Phys. Rev.* B **51** 1803

Landolt M, Allenspach R and Taborelli M 1986 *Surf. Sci.* **178** 311

Lang JK and Baer Y 1979 *Solid State Commun.* **31** 945

Lang JK, Baer Y and Cox PA 1981 *J. Phys. F: Metal Phys.* **11** 121

Lang WC, Padalia BD, Fabian DJ and Watson LM 1974 *J. Electron Spectrosc.* **5** 207

Lapeyre GJ 1969 *Phys. Rev.* **179** 623

Larsson CG 1985 *Surf. Sci.* **152** 213

Laubschat C, Weschke E, Domke M, Simmons CT and Kaindl G 1992 *Surf. Sci.* **270** 605

Laubschat C and Weschke E 1996 *Surf. Rev. Lett.* **3** 1773

Laubschat C 1997 *Appl. Phys.* A **65** 573

Lee BW, Alsenz R, Ignatiev A and Van Hove MA 1978 *Phys. Rev.* B **17** 1510

Lévy JCS 1981 *Surf. Sci. Rep.* **1** 39

Li DQ, Hutchings CW, Dowben PA, Hwang C, Wu RT, Onellion M, Andrews AB and Erskine JL 1991a *J. Magn. Magn. Mat.* **99** 85

Li DQ, Hutchings CW, Dowben PA, Wu RT, Hwang C, Onellion M, Andrews AB and Erskine JL 1991b *J. Appl. Phys.* **70** 6565

Li DQ, Dowben PA and Onellion M 1991c *Bull. Am. Phys. Soc.* **36** 905

Li DQ, Zhang JD, Dowben PA and Onellion M 1992a *Phys. Rev.* B **45** 7272

Li DQ, Zhang JD, Dowben PA, Wu RT and Onellion M 1992b *J. Phys.: Condens. Matter* **4** 3929

Li DQ, Dowben PA and Onellion M 1992c *Mat. Res. Soc. Symp. Proc.* **231** 107

Li DQ, Zhang JD, Dowben PA and Onellion M 1993a *Phys. Rev.* B **48** 5612

Li DQ, Zhang JD, Dowben PA and Garrison K 1993b *J. Phys.: Condens. Matter* **5** L73

Li DQ, Zhang JD, Dowben PA, Garrison K, Johnson PD, Tang H, Walker TG, Hopster H, Scott JC, Weller D and Pappas DP 1993c *Mat. Res. Soc. Symp. Proc.* **313** 451

Li DQ, Dowben PA, Ortega JE and Himpsel FJ 1994 *Phys. Rev.* B **49** 7734

Li DQ, Pearson J, Bader SD, McIlroy DN, Waldfried C and Dowben PA 1995 *Phys. Rev.* B **51** 13895

Li DQ, Pearson J, Bader SD, McIlroy DN, Waldfried C and Dowben PA 1996 *J. Appl. Phys.* **79** 5838

Li H, Tian D, Quinn J, Li YS, Wu SC and Jona F 1992 *Phys. Rev.* B **45** 3853

Li YS, Quinn J, Jona F and Marcus PM 1992 *Phys. Rev.* B **46** 4830

Lindroos M, Barnes CJ, Hu P and King DA 1990 *Chem. Phys. Lett.* **173** 92

Liu LZ, Allen JW, Gunnarsson O, Christensen NE and Andersen OK 1992 *Phys. Rev.* B **45** 8934

Liu SH 1978 in: *Handbook on the Physics and Chemistry of Rare Earths* eds Gschneidner KA and Eyring L (North–Holland, Amsterdam) Vol.1 p.233

Liu SH and Ho KM 1982 *Phys. Rev.* B **26** 7052

Liu SH and Ho KM 1983 *Phys. Rev.* B **28** 4220

Liu SH 1986 *Phys. Rev. Lett.* **57** 269

Liu XM, Wu JX and Zhu JS 1996 *Physica* B **226** 399

Loburets AT, Naumovets AG and Vedula YS 1998 *Surf. Sci.* **399** 297

Loginov MV, Mittsev MA and Pleshkov VA 1992 *Fiz. Tverd. Tela* **34** 3125

Losovyj YB, Medvedev VK, Smereka TP, Palyukh BM and Babkin GV 1982 *Fiz. Tverd. Tela* **24** 2130; *Sov. Phys. Solid State* **24** 1213

Losovyj YB, Medvedev VK, Smereka TP, Babkin GV and Palyukh BM 1984 *Fiz. Tverd. Tela* **26** 1215; *Sov. Phys. Solid State* **26** 738

Losovyj YB, Medvedev VK, Smereka TP, Babkin GV, Palyukh BM and Vasilchishin OS 1986 *Fiz. Tverd. Tela* **28** 3693

Losovyj YB 1997 *Vacuum* **48** 195

Losovyj YB, Dubyk NT and Gonchar FM 1998 *Vacuum* **50** 85

Loucks TL 1966 *Phys. Rev.* **144** 504

Lu ZW, Singh DJ and Krakauer H 1989 *Phys. Rev.* B **39** 4921

Lübcke M, Sonntag B, Niemann W and Rabe P 1986 *Phys. Rev.* B **34** 5184

Lynch DW and Weaver JH 1987 in: *Handbook on the Physics and Chemistry of Rare Earths* eds Gschneidner KA Eyring L and Hüfner S (North–Holland, Amsterdam) Vol.10 p.231
Ma CQ, Krakauer H and Cooper BR 1981 *J. Vac. Sci. Technol.* **18** 581
Macciò M, Pini MG, Trallori L, Politi P and Rettori A 1995 *Phys. Lett.* A **205** 327
Mackintosh AR 1968 *Phys. Lett.* **28A** 217
Mackintosh AR and Anderson OK 1980 in: *Electrons at the Fermi Surface* ed. Springford M (Cambridge University, Cambridge)
Madey TE, Guan J, Nien CH, Dong CZ, Tao HS and Campbell RA 1996 *Surf. Rev. Lett.* **3** 1315
Majkrzak CF, Kwo J, Hong M, Yafet Y, Gibbs D, Chien CL and Bohr J 1991 *Adv. Phys.* **40** 99
Makhov VI and Bondarenko BV 1971 *Soviet Phys. - Solid State* **12** 2986
Malmhäll R, Niihara T, Miyamoto H and Ojima M 1992 *Jpn. J. Appl. Phys.* **31** 1050
Mandl 1992 *Quantum Mechanics* (Wiley, Chichester)
Marquardt DW 1963 *J. Soc. Indust. Appl. Math.* **11** 431
Mårtensson N, Reihl B and Parks RD 1982 *Solid State Commun.* **41** 573
Mårtensson N 1985 *Z. Phys. B – Cond. Matter* **61** 457
Mårtensson N, Stenborg A, Björneholm O, Nilsson A and Andersen JN 1988 *Phys. Rev. Lett.* **60** 1731
Mason MG, Lee ST, Apai G, Davis RF, Shirley DA, Franciosi A and Weaver J H 1981 *Phys. Rev. Lett.* **47** 730
Matsumoto M, Staunton JB and Strange P 1991 *J. Phys.: Condens. Matter* **3** 1453
Matthew JAD 1995 *J. Electron Spectrosc.* **72** 133
McEwen KA and Touborg P 1973 *J. Phys. F: Metal Phys.* **3** 1903
McFeely FR, Kowalczyk SP, Ley L and Shirley DA 1973 *Phys. Lett.* A **45** 227
McFeely FR, Kowalczyk SP, Ley L and Shirley DA 1974 *Phys. Lett.* A **49** 301
McGrath OFK, Ryzhanova N, Lacroix C, Givord D, Fermon C, Miramond C, Saux G, Young S and Vedyayev A 1996 *Phys. Rev.* B **54** 6088
McIlroy DN, Waldfried C, Li DQ, Pearson J, Bader SD, Huang DJ, Johnson PD, Sabiryanov RF, Jaswal SS and Dowben PA 1996 *Phys. Rev. Lett.* **76** 2802
McMasters OD, Holland GE and Gschneidner KA 1978 *J. Cryst. Growth* **43** 577
Medvedev VK, Smereka TP, Stepanovsky SI and Babkin GV 1990 *Ukrain. Fiz. Z.* **35** 251
Medvedev VK, Smereka TP, Stepanovskii SI, Gonchar FM and Kamenetskii RR 1991 *Fiz. Tverd. Tela* **33** 3603
Medvedev VK, Smereka TP, Stepanovsky SI and Gonchar FM 1992 *Ukrain. Fiz. Z.* **37** 1053
Medvedev VK, Smereka TP, Zadorozhnyi LP and Gonchar FM 1993 *Fiz. Tverd. Tela* **35** 1251
Meier R, Weschke E, Bievetski A, Schüssler-Langeheine C, Hu Z and Kaindl G 1998 *Chem. Phys. Lett.* **292** 507
Mélinon P, Fuchs G and Treilleux M 1992 *J. Physique* I **2** 1263
Melmed AJ, Maurice V, Frank O and Block JH 1984 *J. Physique* **45** 47
Meyer RJ, Salaneck WR, Duke CB, Paton A, Griffiths CH, Kovnat L and Meyer LE 1980 *Phys. Rev.* B **21** 4542
Miller A and Dowben PA 1993 *J. Phys.: Condens. Matter* **5** 5459

Mills DL 1971 *Phys. Rev.* B **3** 3887

Min BI, Jansen HJF, Oguchi T and Freeman A J 1986a *J. Magn. Magn. Mat.* **59** 277

Min BI, Jansen HJF, Oguchi T and Freeman AJ 1986b *J. Magn. Magn. Mat.* **61** 139

Min BI, Jansen HJF, Oguchi T and Freeman AJ 1986c *Phys. Rev.* B **34** 369

Min BI, Oguchi T, Jansen HJF and Freeman AJ 1986d *Phys. Rev.* B **34** 654

Mischenko J and Watson PR 1989a *Surf. Sci.* **209** L105

Mischenko J and Watson PR 1989b *Surf. Sci.* **220** L667

Mishra SR, Cummins TR, Waddill GD, Goodman KW, Tobin JG, Gammon WJ, Sherwood T and Pappas DP 1998a *J. Vac. Sci. Technol.* A **16** 1348

Mishra SR, Cummins TR, Waddill GD, Gammon WJ, van der Laan G, Goodman KW and Tobin JG 1998b *Phys. Rev. Lett.* **81** 1306

Mitura Z, Mazurek P, Paprocki K, Mikolajczak P and Beeby JL 1996 *Phys. Rev.* B **53** 10200

Molodtsov SL, Richter M, Danzenbächer S, Wieling S, Steinbeck L and Laubschat C 1997 *Phys. Rev. Lett.* **78** 142

Morishita T, Togami Y and Tsushima K 1985 *J. Phys. Soc. Japan* **54** 37

Mozley S, Nicklin CL, James MA, Steadman P, Norris C and Lohmeier M 1995 *Surf. Sci.* **333** 961

Mühlig A, Günther T, Bauer A, Starke K, Petersen BL and Kaindl G 1998 *Appl. Phys.* A **66** S1195

Mulhollan GA, Garrison K and Erskine JL 1992 *Phys. Rev. Lett.* **69** 3240

Musket RG, McLean W, Colmenares CA, Makowiecki DM and Siekhaus WJ 1982 *Appl. Surf. Sci.* **10** 143

Muthe KP, Gupta MK, Gandhi DP, Vyas JC, Kothiyal GP, Singh KD and Sabharwal SC 1994 *J. Cryst. Growth* **139** 323

Nakamura O, Baba K, Ishii H and Takeda T 1988 *J. Appl. Phys.* **64** 3614

Navas E, Starke K, Laubschat C, Weschke E and Kaindl G 1993 *Phys. Rev.* B **48** 14753

Netzer FP, Wille RA and Grunze M 1981a *Surf. Sci.* **102** 75

Netzer FP, Bertel E and Matthew JAD 1981b *J. Phys. C: Solid State Phys.* **14** 1891

Netzer FP and Bertel E 1982 in: *Handbook on the Physics and Chemistry of Rare Earths* eds Gschneidner KA and Eyring L (North–Holland, Amsterdam) Vol.5 p.217

Netzer FP, Strasser G and Matthew JAD 1983 *Solid State Commun.* **45** 171

Netzer FP, Strasser G, Rosina G and Matthew JAD 1985 *Surf. Sci.* **152–153** 757

Netzer FP and Matthew JAD 1986 *Rep. Prog. Phys.* **49** 621

Netzer FP 1995 *J. Phys.: Condens. Matter* **7** 991

Nicklin CL, Binns C, Norris C, McCluskey P and Barthés-Labrousse MG 1992 *Surf. Sci.* **270** 700

Nicklin CL, Binns C, Norris C, Alleno E and Barthés-Labrousse MG 1994 *Surf. Sci.* **307–309** 858

Nicklin CL, Binns C, Mozley S, Norris C, Alleno E, Barthés-Labrousse MG and van der Laan G 1995 *Phys. Rev.* B **52** 4815

Nicklin CL, Norris C, Steadman P, Taylor JSG and Howes PB 1996 *Physica* B **221** 86

Niemann W, Matzfeldt W, Rabe P, Haensel R and Lübcke M 1987 *Phys. Rev.* B **35** 1099

Nieto JML, Corberan VC and Fierro JLG 1991 *Surf. Int. Anal.* **17** 940

Nigh HE 1963 *J. Appl. Phys.* **34** 3323

Nilsson A, Eriksson B, Mårtensson N, Andersen JN and Onsgaard J 1987 *Phys. Rev.* B **36** 9308

Nilsson A, Eriksson B, Mårtensson N, Andersen JN and Onsgaard J 1988 *Phys. Rev.* B **38** 10357

Nix RM and Lambert RM 1987 *Surf. Sci.* **186** 163

Nix RM, Judd RW and Lambert RM 1988 *Surf. Sci.* **203** 307

Nix RM, Judd RW and Lambert RM 1989a *Surf. Sci.* **215** L316

Nix RM and Lambert RM 1989b *Surf. Sci.* **220** L657

Nolting W, Borstel G, Dambeck T, Fauster T and Vega A 1995 *J. Magn. Magn. Mat.* **140–144** 55

Nyholm R, Chorkendorff I and Schmidt-May J 1984 *Surf. Sci.* **143** 177

Öfner H, Netzer FP and Matthew JAD 1992 *J. Phys.: Condens. Matter* **4** 9795

Ohno TR, Chen Y, Harvey SE, Kroll GH, Benning PJ, Weaver JH, Chibante LPF and Smalley RE 1993 *Phys. Rev.* B **47** 2389

Okane T, Yamada M, Suzuki S, Sato S, Kinoshita T, Kakizaki A, Ishii T, Yuhara J, Katoh M and Morita K 1995 *J. Phys. Soc. Japan* **64** 1673

Okane T, Yamada M, Suzuki S, Sato S, Kinoshita T, Kakizaki A, Ishii T, Kobayashi T, Shimoda S, Iwaki M and Aono M 1996 *J. Electron Spectrosc.* **80** 241

Okane T, Yamada M, Suzuki S, Sato S, Kakizaki A, Kobayashi T, Shimoda S, Iwaki M and Aono M 1998 *J. Phys. Soc. Japan* **67** 264

Olson CG, Wu X, Chen ZL and Lynch DW 1995 *Phys. Rev. Lett.* **74** 992

Onsgaard J, Tougaard S and Morgen P 1979 *Appl. Surf. Sci.* **3** 113

Onsgaard J, Tougaard S, Morgen P and Ryborg F 1980a *J. Electron Spectrosc.* **18** 29

Onsgaard J, Tougaard S, Morgen P and Ryborg F 1980b *Proc. 4th Int. Conf. on Solid Surfaces and 3rd Europ. Conf. on Surf. Sci. Suppl.* **201** 1361

Onsgaard J, Chorkendorff I and Sørensen O 1983 *Phys. Scr.* **T4** 169

Onsgaard J, Chorkendorff I, Ellegaard O and Sørensen O 1984 *Surf. Sci.* **138** 148

Onsgaard J and Chorkendorff I 1986 *Phys. Rev.* B **33** 3503

Ortega JE, Himpsel FJ, Li DQ and Dowben PA 1994 *Solid State Commun.* **91** 807

Ostertag C, Paul J, Cherepkov NA, Oelsner A, Fecher GH and Schönhense G 1997 *Surf. Sci.* **377** 427

Osterwalder J 1985 *Z. Phys. B – Cond. Matter* **61** 113

Over H, Kleinle G, Ertl G, Moritz W, Ernst KH, Wohlgemuth H, Christmann K and Schwarz E 1991 *Surf. Sci.* **254** L469

Pan HB, Xu SH, Xu CS, Lu ED, Xia AD, Zhang GS, Xu PS and Zhang XY 1996 *Chinese Phys. Lett.* **13** 65

Panaccione G, Torelli P, Rossi G, van der Laan G, Sacchi M and Sirotti F 1998 *Phys. Rev.* B **58** R5916

Pang AW, Berger A and Hopster H 1994 *Phys. Rev.* B **50** 6457
Park RL and Madden HH 1968 *Surf. Sci.* **11** 188
Pascal R, Zarnitz C, Bode M and Wiesendanger R 1997a *Phys. Rev.* B **56** 3636
Pascal R, Zarnitz C, Bode M and Wiesendanger R 1997b *Surf. Sci.* **385** L990
Pascal R, Zarnitz C, Bode M and Wiesendanger R 1997c *Appl. Phys.* A **65** 81
Pascal R, Zarnitz C, Bode M, Getzlaff M and Wiesendanger R 1997d *Appl. Phys.* A **65** 603
Pascal R, Zarnitz C, Tödter H, Bode M, Getzlaff M and Wiesendanger R 1998 *Appl. Phys.* A **66** S1121
Paschen U, Sürgers C and Löhneysen Hv 1993 *Z. Phys. B – Cond. Matter* **90** 289
Pasquali L, Fantini P, Nannarone S, Canepa M and Mattera L 1995 *J. Electron Spectrosc.* **76** 133
Patchett AJ, Dhesi SS, Blyth RIR and Barrett SD 1994a *Surf. Sci.* **307–309** 854
Patchett AJ, Dhesi SS, Blyth RIR and Barrett SD 1994b *Surf. Rev. Lett.* **1** 649
Patthey F, Delley B, Schneider WD and Baer Y 1985 *Phys. Rev. Lett.* **55** 1518
Patthey F, Delley B, Schneider WD and Baer Y 1986 *Phys. Rev. Lett.* **57** 270
Patthey F, Bullock EL, Schneider WD and Hulliger F 1993 *Z. Phys. B – Cond. Matter* **93** 71
Paul O, Toscano S, Hürsch W and Landolt M 1990 *J. Magn. Magn. Mat.* **84** L7
Pellissier A, Baptist R and Chauvet G 1989 *Surf. Sci.* **210** 99
Pendry JB 1974 *Low Energy Electron Diffraction* (Academic, London)
Pendry JB 1976 *Surf. Sci.* **57** 679
Pendry JB 1980 *J. Phys. C: Solid State Phys.* **13** 937
Petford-Long AK, Doole RC and Donovan PE 1993 *J. Magn. Magn. Mat.* **126** 41
Pickett WE, Freeman AJ and Koelling DD 1981 *Phys. Rev.* B **23** 1266
Platau A and Karlsson SE 1978 *Phys. Rev.* B **18** 3820
Player MA, Marr GV, Gu E, Savaloni H, Oncan N and Munro IH 1992 *J. Appl. Cryst.* **25** 770
Podloucky R and Glötzel D 1983 *Phys. Rev.* B **27** 3390
Prior KA, Schwawa K and Lambert RM 1978 *Surf. Sci.* **77** 193
Quinn J, Li YS, Jona F and Fort D 1991 *Surf. Sci.* **257** L647
Quinn J, Li YS, Jona F and Fort D 1992 *Phys. Rev.* B **46** 9694
Quinn J, Wang CP, Jona F and Marcus PM 1993 *J. Phys.: Condens. Matter* **5** 541
Raaen S, Braaten NA, Grepstad JK and Qiu SL 1990a *Phys. Scr.* **41** 1001
Raaen S and Braaten NA 1990b *Phys. Rev.* B **41** 12270
Raaen S 1990c *Solid State Commun.* **73** 389
Raaen S, Berg C and Braaten NA 1992a *Surf. Sci.* **270** 953
Raaen S, Berg C and Braaten NA 1992b *Phys. Scr.* **T41** 194
Raiser D and Sens JC 1992 *Appl. Surf. Sci.* **55** 277
Rao GR, Kadowaki Y, Kondoh H and Nozoye H 1995a *Surf. Sci.* **327** 293
Rao GR, Kondoh H and Nozoye H 1995b *Surf. Sci.* **336** 287
Rao GR 1998 *Stud. Surf. Sci. Catalys.* **113** 341
Rau C 1982a *J. Magn. Magn. Mat.* **30** 141
Rau C 1982b *Appl. Surf. Sci.* **13** 310

Rau C 1983 *J. Magn. Magn. Mat.* **31** 874
Rau C and Eichner S 1986a *Phys. Rev.* B **34** 6347
Rau C, Umlauf E and Kuffner H 1986b *Nuc. Instr. Methods Phys. Res.* B **13** 594
Rau C and Robert M 1987 *Phys. Rev. Lett.* **58** 2714
Rau C, Jin C and Robert M 1988a *J. Appl. Phys.* **63** 3667
Rau C and Jin C 1988b *J. Physique* **49** 1627
Rau C, Jin C and Robert M 1989a *Phys. Lett. A* **138** 334
Rau C, Jin C and Liu C 1989b *Vacuum* **39** 129
Rau C and Xing G 1989c *J. Vac. Sci. Technol.* A **7** 1889
Rau C, Jin C and Xing G 1990a *Phys. Lett. A* **144** 406
Rau C, Waters K and Chen N 1990b *J. Electron Spectrosc.* **51** 291
Rau C 1994 *Prog. Surf. Sci.* **46** 135
Reihl B and Himpsel FJ 1982 *Solid State Commun.* **44** 1131
Rettori A, Trallori L, Politi P, Pini MG and Macciò M 1995 *J. Magn. Magn. Mat.* **140–144** 639
Rhee JY, Wang X, Harmon BN and Lynch DW 1995 *Phys. Rev.* B **51** 17390
Richter M and Eschrig H 1989 *Solid State Commun.* **72** 263
Ritley KA and Flynn CP 1998 *Appl. Phys. Lett.* **72** 170
Robinson IK and Tweet DJ 1992 *Rep. Prog. Phys.* **55** 599
Rodbell DS and Moore TW 1964 *Proceedings of the International Conference on Magnetism* (IoP, London) p.427
Roe GM, de Castilho CMC and Lambert RM 1994 *Surf. Sci.* **301** 39
Rosengren A and Johansson B 1982 *Phys. Rev.* B **26** 3068
Rosina G, Bertel E and Netzer FP 1985 *J. Less-Common Metals* **111** 285
Rosina G, Bertel E, Netzer FP and Redinger J 1986a *Phys. Rev.* B **33** 2364
Rosina G, Bertel E and Netzer FP 1986b *Phys. Rev.* B **34** 5746
Rossi G and Barski A 1986 *Solid State Commun.* **57** 277
Rous PJ and Pendry JB 1989a *Surf. Sci.* **219** 355
Rous PJ and Pendry JB 1989b *Surf. Sci.* **219** 373
Rous PJ, Van Hove MA and Somorjai GA 1990 *Surf. Sci.* **226** 15
Rous PJ 1993 *Surf. Sci.* **296** 358
Rous PJ 1994 *J. Phys.: Condens. Matter* **6** 8103
Roustila A, Severac C, Chene J and Percheron-Guégan A 1994 *Surf. Sci.* **311** 33
Sabiryanov RF and Jaswal SS 1997 *Phys. Rev.* B **55** 4117
Sacchi M, Sakho O, Sirotti F, Jin X and Rossi G 1991 *Surf. Sci.* **251** 346
Salas FH and Mirabal-Garciá M 1990 *Phys. Rev.* B **41** 10859
Salas FH 1991 *J. Phys.: Condens. Matter* **3** 2839
Salas FH and Weller D 1993 *J. Magn. Magn. Mat.* **128** 209
Savaloni H, Gu E, Player MA, Marr GV and Munro IH 1992a *Rev. Sci. Instrum.* **63** 1494
Savaloni H, Player MA, Gu E and Marr GV 1992b *Vacuum* **43** 965
Savaloni H, Player MA, Gu E and Marr GV 1992c *Rev. Sci. Instrum.* **63** 1497

Savaloni H and Player MA 1995 *Thin Solid Films* **256** 48

Scarfe JA, Law AR, Hughes HP, Bland JAC, Roe GM and Walker AP 1992 *Phys. Stat. Sol.* B **171** 377

Schieber M 1967 in: *Crystal Growth* ed. Perser HS (Pergamon, New York) p.271

Schiller R, Müller W and Nolting W 1997 *J. Magn. Magn. Mat.* **169** 39

Schirber JE, Schmidt FA, Harmon BN and Koelling DD 1976 *Phys. Rev. Lett.* **36** 448

Schirber JE, Switendick AC and Schmidt FA 1983 *Phys. Rev.* B **27** 6475

Schneider WD, Laubschat C and Reihl B 1983 *Phys. Rev.* B **27** 6538

Schneider WD, Delley B, Wuilloud E, Imer JM and Baer Y 1985 *Phys. Rev.* B **32** 6819

Schneider WD, Gantz T, Richter M, Molodtsov SL, Boysen J, Engelmann P, Segovia-Cabrero P and Laubschat C 1997 *Surf. Sci.* **377** 275

Schneider W, Molodtsov SL, Richter M, Gantz T, Engelmann P and Laubschat C 1998 *Phys. Rev.* B **57** 14930

Scholz B, Brand RA and Keune W 1994 *Phys. Rev.* B **50** 2537

Schorsch V, Beaurepaire E, Barbier A, Deville JP, Carrière B, Gourieux T and Hricovini K 1994 *Surf. Sci.* **309** 603

Schreifels JA, Deffeyes JE, Neff LD and White JM 1982 *J. Electron Spectrosc.* **25** 191

Schwarz RB and Johnson WL 1983 *Phys. Rev. Lett.* **51** 415

Searle C, Blyth RIR, White RG, Tucker NP, Lee MH and Barrett SD 1995 *J. Synchrotron Rad.* **2** 312

Shakirova SA and Shevchenko MA 1990 *Fiz. Tverd. Tela* **32** 688

Shakirova SA, Pleshkov VA and Rump GA 1992 *Surf. Sci.* **279** 113

Shaw EA, Ormerod RM and Lambert RM 1992 *Surf. Sci.* **275** 157

Sherwood TS, Mishra SR, Popov AP and Pappas DP 1998 *J. Vac. Sci. Technol.* A **16** 1364

Shi NL and Fort D 1987 *Chinese J. Met. Sci. Technol.* **3** 156

Shih H, Jona F, Jepsen D and Marcus P 1976a *J. Phys. C: Solid State Phys.* **9** 1405

Shih H, Jona F, Jepsen D and Marcus P 1976b *Commun. Phys.* **1** 25

Shikin AM, Prudnikova GV, Fedorov AV and Adamchuk VK 1994 *Surf. Sci.* **309** 205

Shikin AM, Prudnikova GV, Adamchuk VK, Molodtsov SL, Laubschat C and Kaindl G 1995 *Surf. Sci.* **331–333** 517

Shimizu T, Nonaka H and Arai K 1992 *Surf. Int. Anal.* **19** 365

Shirley DA 1978 in: *Photoemission in Solids I* eds Cardona M and Ley L (Springer–Verlag, Berlin) ch.4

Siegmann HC 1992 *J. Phys.: Condens. Matter* **4** 8395

Simonson RJ, Wang JR and Ceyer ST 1987 *J. Phys. Chem.* **91** 5681

Singh DJ 1991a *Phys. Rev.* B **43** 6388

Singh DJ 1991b *Phys. Rev.* B **44** 7451

Singh P, Mandale AB and Badrinarayanan S 1988 *J. Less-Common Metals* **141** 1

Skomski R, Waldfried C and Dowben PA 1998 *J. Phys.: Condens. Matter* **10** 5833

Skriver HL 1982 in: *Systematics and Properties of the Lanthanides* ed. Sinha SP (Reidel, Dordrecht)

Smereka TP, Ubogyi IM and Losovyi YB 1995 *Vacuum* **46** 429
Smith NV 1988 *Rep. Prog. Phys.* **51** 1227
Sokolov J, Quinn J, Jona F and Marcus PM 1989 *Bull. Am. Phys. Soc.* **34** 579
Sørland GH and Raaen S 1993 *Physica* B **183** 415
Spanjaard D, Guillot C, Desjonquères MC, Tréglia G and Lecante J 1985 *Surf. Sci. Rep.* **5** 1
Spedding FH 1970 in: *Handbook of Chemistry and Physics* (Chemical Rubber Co., Cleveland Ohio) p.B-253
Spedding FH, Cress D and Beaudry BJ 1971 *J. Less-Common Metals* **23** 263
Spedding FH and Croat JJ 1973 *J. Chem. Phys.* **58** 5514
Speier W, Fuggle JC, Zeller R, Ackermann B, Szot K, Hillebrecht FU and Campagna M 1984 *Phys. Rev.* B **30** 6921
Starke K, Navas E, Baumgarten L and Kaindl G 1993 *Phys. Rev.* B **48** 1329
Starke K, Baumgarten L, Arenholz E, Navas E and Kaindl G 1994 *Phys. Rev.* B **50** 1317
Starke K, Navas E, Arenholz E, Baumgarten L and Kaindl G 1995a *Appl. Phys.* A **60** 179
Starke K, Navas E, Arenholz E, Baumgarten L and Kaindl G 1995b *IEEE Trans. Mag.* **31** 3313
Stassis C, Smith GS, Harmon BN, Ho KM and Chen Y 1985 *Phys. Rev.* B **31** 6298
Stenborg A and Bauer E 1987a *Phys. Rev.* B **36** 5840
Stenborg A and Bauer E 1987b *Surf. Sci.* **185** 394
Stenborg A and Bauer E 1987c *Surf. Sci.* **189** 570
Stenborg A and Bauer E 1988 *Solid State Commun.* **66** 561
Stenborg A, Andersen JN, Björneholm O, Nilsson A and Mårtensson N 1989a *Phys. Rev. Lett.* **63** 187
Stenborg A, Björneholm O, Nilsson A, Mårtensson N, Andersen JN and Wigren C 1989b *Phys. Rev.* B **40** 5916
Stenborg A, Björneholm O, Nilsson A, Mårtensson N, Andersen JN and Wigren C 1989c *Surf. Sci.* **211** 470
Stetter U, Farle M, Baberschke K and Clark WG 1992a *Phys. Rev.* B **45** 503
Stetter U, Aspelmeier A and Baberschke K 1992b *J. Magn. Magn. Mat.* **117** 183
Sticht J and Kübler J 1985 *Solid State Commun.* **53** 529
Stranski JN and Krastanov L 1938 *Ber. Akad. Wiss. Wien.* **146** 797
Strasser G, Bertel E and Netzer FP 1983 *J. Catalysis* **79** 420
Strasser G, Rosina G, Bertel E and Netzer FP 1985 *Surf. Sci.* **152–153** 765
Strisland F and Raaen S 1996 *J. Electron Spectrosc.* **77** 25
Strisland F, Raaen S, Ramstad A and Berg C 1997 *Phys. Rev.* B **55** 1391
Strisland F, Ramstad A, Berg C and Raaen S 1998 *Surf. Sci.* **410** 344
Strozier J and Jones R 1971 *Phys. Rev.* B **3** 3228
Surgers C and Vonlohneysen H 1992 *Thin Solid Films* **219** 69
Surplice NA and Brearley W 1978 *Surf. Sci.* **72** 84
Szotek Z, Temmerman WM and Winter H 1991a Physica B **165** 275
Szotek Z, Temmerman WM and Winter H 1991b Physica B **172** 19
Taborelli M, Allenspach R, Boffa G and Landolt M 1986a *Phys. Rev. Lett.* **56** 2869

Taborelli M, Allenspach R and Landolt M 1986b *Phys. Rev.* B **34** 6112

Takakuwa Y, Takahashi S, Suzuki S, Kono S, Yokotsuka T, Takahashi T and Sagawa T 1982 *J. Phys. Soc. Japan* **51** 2045

Takakuwa Y, Suzuki S, Yokotsuka T and Sagawa T 1984 *J. Phys. Soc. Japan* **53** 687

Takeda K, Ishi H and Nakamura O 1989 *Jpn. Kokai Tokkyo Koho JP* 89 122 105 patent

Tanaka Y, Kamei M and Gotoh Y 1995a *Surf. Sci.* **336** 13

Tanaka Y, Kamei M and Gotoh Y 1995b *Jpn. J. Appl. Phys.* **34** 5774

Tanaka Y, Kamei M and Gotoh Y 1996a *Surf. Sci.* **360** 74

Tanaka Y, Kamei M and Gotoh Y 1996b *J. Cryst. Growth* **169** 299

Tang CC, Stirling WG, Jones DL, Wilson CC, Haycock PW, Rollason AJ, Thomas AH and Fort D 1992 *J. Magn. Magn. Mat.* **103** 86

Tang H, Weller D, Walker TG, Scott JC, Chappert C, Hopster H, Pang AW, Dessau DS and Pappas DP 1993a *Phys. Rev. Lett.* **71** 444

Tang H, Walker TG, Hopster H, Pappas DP, Weller D and Scott JC 1993b *Phys. Rev.* B **47** 5047

Tang H, Walker TG, Hopster H, Pappas DP, Weller D and Scott JC 1993c *J. Magn. Magn. Mat.* **121** 205

Tang H, Walker TG, Hopster H, Pang AW, Weller D, Scott JC, Chappert C, Pappas DP and Dessau DS 1993d *J. Appl. Phys.* **73** 6769

Tang J, Lawrence JM and Hemminger JC 1993a *Phys. Rev.* B **47** 16477

Tang J, Lawrence JM and Hemminger JC 1993b *Phys. Rev.* B **48** 15342

Tanuma S, Powell CJ and Penn DR 1991 *Surf. Int. Anal.* **17** 911

Tao L, Goering E, Horn S and denBoer ML 1993 *Phys. Rev.* B **48** 15289

Theis-Bröhl K, Ritley KA, Flynn CP, Hamacher K, Kaiser H and Rhyne JJ 1996 *J. Appl. Phys.* **79** 4779

Theis-Bröhl K, Ritley KA, Flynn CP, Hamacher K, Kaiser H and Rhyne JJ 1997a *J. Appl. Phys.* **81** 5375

Theis-Bröhl K, Ritley KA, Flynn CP, Van Nostrand JE, Cahill DG, Hamacher K, Kaiser H and Rhyne JJ 1997b *J. Magn. Magn. Mat.* **166** 27

Thole BT, Wang XD, Harmon BN, Li DQ and Dowben PA 1993 *Phys. Rev.* B **47** 9098

Thromat N, Gautiersoyer M and Bordier G 1996 *Surf. Sci.* **345** 290

Tibbetts GG and Egelhoff WF 1980 *J. Vac. Sci. Technol.* **17** 458

Titterington DJ and Kinniburgh CG 1980 *Computer Phys. Commun.* **20** 237

Tober ED, Ynzunza RX, Westphal C and Fadley CS 1996 *Phys. Rev.* B **53** 5444

Tober ED, Ynzunza RX, Palomares FJ, Wang Z, Hussain Z, Van Hove MA and Fadley CS 1997 *Phys. Rev. Lett.* **79** 2085

Tober ED, Palomares FJ, Ynzunza RX, Denecke R, Morais J, Wang Z, Bino G, Liesegang J, Hussain Z and Fadley CS 1998 *Phys. Rev. Lett.* **81** 2360

Tougaard S and Ignatiev A 1981 *Surf. Int. Anal.* **3** 3

Tougaard S and Ignatiev A 1982 *Surf. Sci.* **115** 270

Tougaard S and Hansen HS 1990 *Surf. Sci.* **236** 271

Trappmann T, Gajdzik M, Sürgers C and von Löhneysen Hv 1997 *Europhys. Lett.* **39** 159

Tsui F, Flynn CP, Salamon MB, Erwin RW, Borchers JA and Rhyne JJ 1991 *Phys. Rev.* B **43** 13320

Tucker NP, Blyth RIR, White RG, Lee MH, Robinson AW and Barrett SD 1995 *J. Synchrotron Rad.* **2** 252

Tucker NP, Blyth RIR, White RG, Lee MH, Searle C and Barrett SD 1998 *J. Phys.: Condens. Matter* **10** 6677

Ufuktepe Y 1993 *J. Phys.: Condens. Matter* **5** L213

Unertl WN and McKay SR 1984 in: *Determination of Surface Structure by LEED* eds Marcus PM and Jona F (Plenum, New York) p.261

Unertl WN and Thapliyal H 1975 *J. Vac. Sci. Technol.* **12** 263

van der Laan G, Arenholz E, Navas E, Bauer A and Kaindl G 1996 *Phys. Rev.* B **53** R5998

van der Veen JF, Himpsel FJ and Eastman DE 1980 *Phys. Rev. Lett.* **44** 189

Van Hove MA, Tong SY and Elconin MH 1977 *Surf. Sci.* **64** 85

Van Hove MA and Tong SY 1979 *Surface Crystallography by LEED* (Springer, Berlin)

Van Hove MA, Moritz W, Over H, Rous PJ, Wander A, Barbieri A, Meterer N, Starke U and Somorjai GA 1993 *Surf. Sci. Rep.* **19** 191

Vaterlaus A, Beutler T and Meier F 1991 *Phys. Rev. Lett.* **67** 3314

Venvik HJ, Berg C, Borg A and Raaen S 1996 *Phys. Rev.* B **53** 16587

Vescovo E, Rochow R, Kachel T and Carbone C 1992 *Phys. Rev.* B **46** 4788

Vescovo E, Rader O, Kachel T, Alkemper U and Carbone C 1993a *Phys. Rev.* B **47** 13899

Vescovo E, Carbone C and Rader O 1993b *Phys. Rev.* B **48** 7731

Vescovo E and Carbone C 1996 *Phys. Rev.* B **53** 4142

Vlieg E, Lohmeier M and van der Vegt HA 1995 *Nucl. Instr. Meth. Phys. Res.* B **97** 358

Vogel J, Sacchi M, Sirotti F and Rossi G 1993 *Appl. Surf. Sci.* **65–66** 170

Vogel J and Sacchi M 1994 *J. Electron Spectrosc.* **67** 181

Vogel J and Sacchi M 1996 *Surf. Sci.* **365** 831

Volmer M and Weber A 1926 *Z. Phys. Chem.* **119** 277

Waldfried C, McIlroy DN, Hutchings CW and Dowben PA 1996 *Phys. Rev.* B **54** 16460

Waldfried C, McIlroy DN and Dowben PA 1997 *J. Phys.: Condens. Matter* **9** 10615

Waldfried C, Zeybek O, Bertrams T, Barrett SD and Dowben PA 1998 *Mat. Res. Soc. Symp. Proc.* **528** 147

Waldrop JR, Grant RW, Wang YC and Davis RF 1992 *J. Appl. Phys.* **72** 4757

Walker AP and Lambert RM 1992 *J. Phys. Chem.* **96** 2265

Wang XD, Xu QK, Hashizume T, Shinohara H, Nishina Y and Sakurai T 1994 *Appl. Surf. Sci.* **76** 329

Warren JP, Zhang X, Andersen JET and Lambert RM 1993 *Surf. Sci.* **287** 222

Watson RE, Freeman AJ and Dimmock JO 1968 *Phys. Rev.* **167** 497

Weller D, Alvarado SF, Gudat W, Schröder K and Campagna M 1985a *Phys. Rev. Lett.* **54** 1555

Weller D and Alvarado SF 1985b *Z. Phys. B – Cond. Matter* **58** 261

Weller D, Alvarado SF and Campagna M 1985c *Physica* B **130** 72

Weller D, Alvarado SF, Campagna M, Gudat W and Sarma DD 1985d *J. Less-Common Metals* **111** 277

Weller D and Alvarado SF 1986a *J. Appl. Phys.* **59** 2908

Weller D and Sarma DD 1986b *Surf. Sci.* **171** L425

Weller D and Alvarado SF 1988 *Phys. Rev.* B **37** 9911

Welz M, Moritz W and Wolf D 1983 *Surf. Sci.* **125** 473

Wertheim GK and Campagna M 1977 *Chem. Phys. Lett.* **47** 182

Wertheim GK and Crecelius G 1978 *Phys. Rev. Lett.* **40** 813

Weschke E, Laubschat C, Simmons T, Domke M, Strebel O and Kaindl G 1991 *Phys. Rev.* B **44** 8304

Weschke E, Laubschat C, Höhr A, Starke K, Navas E, Baumgarten L, Fedorov AV and Kaindl G 1994 *J. Electron Spectrosc.* **68** 515

Weschke E and Kaindl G 1995a *J. Electron Spectrosc.* **75** 233

Weschke E, Höhr A, Vandré S, Schüssler-Langeheine C, Bodker F and Kaindl G 1995b *J. Electron Spectrosc.* **76** 571

Weschke E, Schüssler-Langeheine C, Meier R, Fedorov AV, Starke K, Hübinger F and Kaindl G 1996 *Phys. Rev. Lett.* **77** 3415

Weschke E, Schüssler-Langeheine C, Meier R, Fedorov AV, Starke K, Hübinger F and Kaindl G 1997a *Surf. Sci.* **377** 487

Weschke E, Schüssler-Langeheine C, Meier R, Kaindl G, Sutter C, Abernathy D and Grübel G 1997b *Phys. Rev. Lett.* **79** 3954

Weschke E, Höhr A, Kaindl G, Molodtsov SL, Danzenbächer S, Richter M and Laubschat C 1998 *Phys. Rev.* B **58** 3682

White RG, Blyth RIR, Tucker NP, Lee MH and Barrett SD 1995 *J. Synchrotron Rad.* **2** 261

White RG 1996 PhD thesis, University of Liverpool, UK

White RG, Murray PW, Lee MH, Tucker NP and Barrett SD 1997 *Phys. Rev.* B **56** R10071

Wieliczka DM, Weaver JH, Lynch DW and Olson CG 1982 *Phys. Rev.* B **26** 7056

Wieliczka DM, Olson CG and Lynch DW 1984 *Phys. Rev. Lett.* **52** 2180

Wieliczka DM and Olson CG 1990 *J. Vac. Sci. Technol.* A **8** 891

Wieling S, Molodtsov SL, Gantz T, Hinarejos JJ, Laubschat C and Richter M 1998 *Phys. Rev.* B **58** 13219

Wiesendanger R 1994 *Scanning Probe Microscopy and Spectroscopy: Methods and Applications* (Cambridge University, Cambridge)

Wiesendanger R, Bode M, Dombrowski R, Getzlaff M, Morgenstern M and Wittneven C 1998 *Jpn. J. Appl. Phys.* **37** 3769

Williams RW, Loucks TL and Mackintosh AR 1966 *Phys. Rev. Lett.* **16** 168

Williams RW and Mackintosh AR 1968 *Phys. Rev.* **168** 679

Winiarski A 1982 *J. Cryst. Growth* **57** 443

Witkowski N, Bertran F, Gourieux T, Kierren B, Malterre D and Panaccione G 1997 *Phys. Rev.* B **56** 12054

Wolf SA, Qadri SB, Claassen JH, Francavilla TL and Dalrymple BJ 1986 *J. Vac. Sci. Technol.* A **4** 524

Wood EA 1964 *J. Appl. Phys.* **35** 1306

Woodruff DP and Delchar TA 1994 *Modern Techniques of Surface Science* (Cambridge University, Cambridge)

Wu RQ, Li C, Freeman AJ and Fu CL 1991a *Phys. Rev.* B **44** 9400

Wu RQ and Freeman AJ 1991b *J. Magn. Magn. Mat.* **99** 81

Wu SC, Li H, Tian D, Quinn J, Li YS, Jona F, Sokolov J and Christensen NE 1990 *Phys. Rev.* B **41** 11911; ibid. **43** 12060

Wu SC, Li H, Li YS, Tian D, Quinn J, Jona F and Fort D 1991 *Phys. Rev.* B **44** 13720

Wu SC, Li H, Li YS, Tian D, Quinn J, Jona F, Fort D and Christensen NE 1992 *Phys. Rev.* B **45** 8867

Wuilloud E, Moser HR, Schneider WD and Baer Y 1983 *Phys. Rev.* B **28** 7354

Yang KY, Homma H and Schuller IK 1988 *J. Appl. Phys.* **63** 4066

Yeh JJ and Lindau I 1985 *At. Data Nucl. Data Tables* **32** 1

Zadorozhnyi LP, Medvedev VK, Smereka TP and Gonchar FM 1992 *Fiz. Tverd. Tela* **34** 1051

Zagwijn PM, Frenken JWM, van Slooten U and Duine PA 1997 *Appl. Surf. Sci.* **111** 35

Zanazzi E and Jona F 1977 *Surf. Sci.* **62** 61

Zangwill A 1992 *Physics at Surfaces* (Cambridge University, Cambridge)

Zhang F, Thevuthasan S, Scalettar RT, Singh RRP and Fadley CS 1995 *Phys. Rev.* B **51** 12468

Zhang FP, Xu PS, Xu SH, Lu ED, Yu XJ and Zhang XY 1997 *Chinese Phys. Lett.* **14** 553

Zhang H, Satija SK, Gallagher PD, Dura JA, Ritley K, Flynn CP and Ankner JF 1995 *Phys. Rev.* B **52** 17501

Zhang H, Satija SK, Gallagher PD, Dura JA, Ritley K, Flynn CP and Ankner JF 1996 *Physica* B **221** 450

Zhang JD, Dowben PA, Li DQ and Onellion M 1995 *Surf. Sci.* **329** 177

Zhang XX, Ferrater C, Zquiak R and Tejada J 1994 *IEEE Trans. Mag.* **30** 818

Zimmer RS and Robertson WD 1974 *Surf. Sci.* **43** 61

Zuo JK, Wendelken JF, Dürr H and Liu CL 1994 *Phys. Rev. Lett.* **72** 3064

INDEX